Design of Pressure Vessels

Design of Pressure Vessels

Edited By

Subhash Reddy Gaddam

CRC Press is an imprint of the
Taylor & Francis Group, an **informa** business

First edition published 2021
by CRC Press
6000 Broken Sound Parkway NW, Suite 300, Boca Raton, FL 33487-2742

and by CRC Press
2 Park Square, Milton Park, Abingdon, Oxon, OX14 4RN

CRC Press is an imprint of Taylor & Francis Group, LLC

Library of Congress Cataloging-in-Publication Data
Names: Gaddam, Subhash Reddy, author.
Title: Design of pressure vessels / Subhash Reddy Gaddam.
Description: First edition. | Boca Raton, FL : CRC Press/Taylor & Francis Group, LLC, 2021. | Includes bibliographical references and index.
Identifiers: LCCN 2020035117 (print) | LCCN 2020035118 (ebook) | ISBN 9780367550646 (hardback) | ISBN 9781003091806 (ebook)
Subjects: LCSH: Pressure vessels--Design and construction.
Classification: LCC TA660.T34 G34 2021 (print) | LCC TA660.T34 (ebook) | DDC 681/.76041--dc23
LC record available at https://lccn.loc.gov/2020035117
LC ebook record available at https://lccn.loc.gov/2020035118

ISBN: 978-0-367-55064-6 (hbk)
ISBN: 978-0-367-55066-0 (pbk)
ISBN: 978-1-003-09180-6 (ebk)

DOI: 10.1201/9781003091806

Typeset in Times
by SPi Global, India

Contents

Preface

On the subject of pressure vessel design, several books and codes are available. They are voluminous and expensive. Some of the books contain derivations of equations by integration and others ready to use formulas. Codes contain equations by rule.

In this generation with availability of software programs even for simple calculations, the requirement of fundamental stress analysts is limited to the development of computer software programs for analytical and numerical engineering stress analysis. Most of the design engineers are becoming software operators.

But it is the belief of author (who worked in this field from the slide rule generation to PVEllite and Ceasar generation) that today's designers shall have sufficient basic knowledge of the subject to effectively use the software programs and codes to give optimum design. He shall be in a position to comment on the software for more effectiveness and suggest changes in codes. For designers, the design software shall be only a tool to reduce time and accuracy.

Efforts are made in this book to fill this gap. This book avoids lengthy integration which is generally skipped by present engineers and explains the equations by simple fundamental design methods. It will give feeling of how the parts will resist and deform under various types of loads and explains various tools of design from fundamentals giving practical applications so as to modify or combine them where direct formulas are not available.

Effort of this book is not to develop knowledge parallel to software, but to take stress analysis software operating engineers to a step inside, backend, to understand the loading diagram or free body diagram, deformation etc., so that they can be confident in selecting the correct program, giving error free input, analyzing the report and validating the results. Guidelines are provided for fundamental stress analysis.

This book will be useful to new graduates and post-graduates for the subject of solid mechanics and aspiring career in pressure vessel design field, advance education, and research and to working professionals in static equipment, pressure vessels, heat exchangers, boilers, and similar equipments.

Efforts are made to include an optimum material so that the reader will not skip even a small part of the chapter or section of the chapter interested or relevant to him/her. The pressure-retaining parts and connected non-pressure parts in general are many types. Efforts are made to cover design of all such parts, but in depth analysis is covered to parts related to the boiler industry. This book contains terms, and language used by working professionals may be marginally different from that used by faculty or is usually found in most of the books.

This book would not have been possible without the support of several friends and Thermal systems (Hyderabad) Pvt. Ltd where the author works. The author is thankful to all of them particularly, SimharajuYogesh, Toji Tharakan, and Lakshman for providing figures and Praveen Mittapelli, Mani Kumar, Ram Sri Pavan, Mastan, and Dr. Raman Goud for their support.

Preface

Author

Subhash Reddy Gaddam has been associated with design, development, engineering, project execution, manufacturing, and commissioning of pressure vessels and boilers of various types and capacities for over 50 years.

He is a 1970 graduate in Mechanical Engineering from Osmania University, College of Engineering. Starting as a graduate trainee with the Indian subsidiary of the Babcock & Wilcox Ltd., U.K., where he worked on the shop floor and construction site in the commissioning of a 120MW boiler, he spent over a decade in design and engineering of industrial and power boilers at various large industries in India. He then joined the Andhra Pradesh State Government as Inspector of Boilers until his retirement as the Director and Head of the Department of Boilers in 2006. He is currently an adviser for mechanical design of pressure vessels, heat exchangers and waste heat boilers at Thermal Systems (Hyderabad) Pvt. Ltd.

This book has been painstakingly compiled over the past several years and contains deep technical insights from his experience spanning 50 years. He has personally developed worksheets for design calculations of all components of pressure vessels & boilers described in this book.

1 Introduction

Mechanical design (analysis) reported in this book is the calculation of thickness for maximum allowable stress or maximum induced relevant stresses for a given thickness of any element for which equation is derived using maximum principal stress theory of failure by any analytical method of analysis and relevant design theories. The activities involved in connection with pressure vessel design for manufacturing them are briefly covered in this chapter.

Design in general starts with process design, which is normally performed in R & D organization. The next step is plant design for satisfactory operation and facilitating maintenance of the plant, which is the design and linking of other auxiliary parts:

a. Equipments in which process or part of it takes place.
b. Controlling parts such as instruments, valves, etc., which are used to isolate or regulate the process parameters such as flow, pressure, and temperature or any other required controls to perform.
c. Instruments, gauges, control valves, etc., which are involved in measuring the process parameters in field as well as in a control room/a digital control system.
d. Auxiliary rotating equipments such as pumps, fans, and other such equipments.

Linking between equipments is performed by process piping with required instruments, field gauges, process valves, etc. as per piping and instrumentation diagrams required for the process.

Instruments and valves are normally manufactured by specialized organizations. Other auxiliary equipments are also manufactured by specialist organizations mostly involving casting, forging, and fabrication. Equipment and piping are manufactured by fabricating mostly welding using raw materials such as plates, tubes, pipes, pipe fittings, etc.

Equipment, piping, and most of the line/process instruments are under pressure and hence are called pressure vessels.

Pressure vessels are first designed (detailed process design and operating parameters involving heat transfer and fluid dynamics) to give arrangement, shape, and size required for a specified process. Then, design data are obtained, residual process design is carried out, materials are selected, remaining geometry and arrangement are finalized, and thickness is calculated to withstand prevailing loads.

The pressure vessel consists mostly of plates and tubular shaped parts formed as required and welded. Most of the components of pressure vessels excluding nonpressure parts such as attachment supports and internals are subjected to pressure and

DOI: 10.1201/9781003091806-1

hence called pressure parts. Selection of materials, calculation of thickness, and other requirements shall be in accordance with relevant national technical codes and statutory rules/acts. Materials for nonpressure parts welded to pressure parts shall be the same as those for pressure parts. All the nonpressure parts are to be designed as per relevant standards, fundamentals, and accepted technical literature and practices. It also includes the preparation of detailed specification or data sheets of controls, instruments, and auxiliaries for procurement/manufacturing. Design also includes arrangement and calculation of handling and shipping components.

Design details not given in codes for pressure parts are submitted for the acceptance of the statutory or authorized inspector, and the designer provides details of design from other sources such as other codes, standards, fundamentals, and excepted practices, which will be as safe as those provided by the rules of code. Alternately, strength of pressure parts can be established by a proof test or finite element analysis.

The next step is engineering and producing drawings for arrangement and details of parts of equipments required for manufacturing, erection, and shipping. The material list, detailed data sheets for the purchase of raw materials, and other components are prepared.

Pressure vessels including boilers are normally allowed to use in most of the countries duly registered, and certified for use similar to automobile vehicles. The statutory and/or mandatory rules vary from country to country and are not the same for boilers and pressure vessels and more stringent for boilers. In general, complete activity of manufacturing boilers and pressure vessels (generally termed pressure parts) right from raw materials until completion of installation at the place of their use is inspected as per the Acts and Rules of the country by agency authorized in the Acts and Rules for the purpose before being registered and certified for use.

The manufacturing drawings are to be approved by the inspection agency duly conforming to the rules (code), the components are inspected, required tests are carried out in various stages, identification marks are stamped, and finally the pressure vessels are hydraulically tested for preliminary check of their resistance to pressure before transporting to the place of erection. The code stipulates that the manufacturer should prepare all required documents duly signed by them and the inspection agency which will accompany the pressure component for verification before erection and for registration.

2 Material

2.1 METALLURGICAL FUNDAMENTALS

- *Toughness*: Strength and ductility = area under stress-strain curve, toughness of structural steel > spring steel.
- *Microstructure*: Ferrite, austenite, cementite, pearlite, and martensite.
- *Ferrite*: It has a body centered cubic crystal structure. It is pure or 0.01% carbon (dissolve at room temperature), soft, and ductile.
- *Austenite*: It has a face centered cubic crystal structure and exists over 912°C.
- *Cementite*: Compounds of iron and carbon iron carbide (Fe_3C) are hard and brittle.
- *Pearlite*: The lamented structure is formed of alternate layers of ferrite and cementite. It combines the hardness and strength of cementite with the ductility of ferrite. The laminar structure acts as a barrier to crack movement as in composite, which gives its toughness.
- *Martensite*: It has a very hard needle-like structure of iron and carbon and is formed by rapid cooling of the austenitic structure (above the upper critical temperature), which needs to be modified by tempering before acceptable properties are reached.

2.2 VARIETY OF MATERIALS

The following materials are normally used in manufacture of boilers and pressure vessels.

2.2.1 Carbon Steel (CS)

CS generally contains carbon 0.3% max. CS is easily formable, weldable, and machinable. However, CS has limited corrosion resistance and low creep strength and is normally not advised for use in the creep range that is for higher temperatures more than 400°C.

2.2.2 Low Alloy Steel (LAS)

LAS contains one or more alloying elements to improve mechanical and other properties such as corrosion resistance, strength at elevated temperatures, impact and toughness at low temperatures, and resistance to pitting. The most usual used

DOI: 10.1201/9781003091806-2

alloying elements are chromium 9% max, molybdenum 1% max, traces of vanadium, nickel, etc. Nickel up to 9% is used for special requirements. Objects of main alloying elements are given below:

- *Chromium* (Cr) improves resistance to oxidation and improves corrosion resistance, abrasion resistance, high-temperature strength, and resistance to high-pressure hydrogen.
- *Molybdenum* (Mo) improves strength at elevated temperature, creep resistance, and resistance to pitting.
- *Vanadium* (V) improves wear resistance, high-temperature strength, and resistance to high-pressure hydrogen.
- *Nickel* (Ni) improves toughness, low temperature properties, and corrosion resistance.

2.2.3 Stainless Steel (SS)

There are three varieties of SS: austenitic, martensitic, and ferritic.

1. *Austenitic* is corrosion resistant, tough, ductile, and easy to form and weld. These steels are suitable for low service temperature in the range of – 80 to 250°C as well as high service temperature in the range of 400 to 500°C. Some special grades such as 304H, 316H, and 321H are suitable for very high service temperatures in the range of 500 to 816°C.
2. *Martensitic* contains 1 to 14% chromium. Type 409, 410, and 410S are commonly used for noncooled and nonpressure parts (PPs) required for high-temperature service.
3. *Ferritic* contains 14 to 20% chromium. Type 429, 430, and 439 belong to this category. These grades are corrosion-resistant and suitable for higher service temperatures up to 650°C and up to 800°C for noncooled and non-PPs.

2.2.4 Nonferrous Alloys

Inconel and incoloy are used for very high temperature services above 800°C, for noncooled and non-PPs such as coil supports, tube supports in high temperature radiation zones, and furnace components.

1. *Inconel*: Different grades of inconel contain nickel in the range of 50 to 75% and chromium in the range of 15 to 25% and are suitable for high temperature above 800°C. SB-168 Alloy Nos. 6600, 6601, 6617, and 6690 belong to this group.
2. *Incoloy*: Different grades of incoloy contain nickel in the range of 30 to 40% and chromium around 20% and are suitable for high temperature above 800°C. SB-409 Alloy No.8120, 8800, 8810, and 8811 belong to this group.

Other alloys are used as required to resist the various chemical effects of different fluids in contact.

2.2.5 Nonmetallic Materials

- Insulation
- Refractory

2.3 MATERIAL SELECTION

Generally, codes for pressure vessels do not contain material selection. Codes stipulate only the manufacturing process, chemical composition, mechanical properties (mandatory and optional), inspection, heat treatment and testing requirements, certification, marking etc., for the general or listed materials. Most of the nations have their own material standards to suit various processes and functional requirements. Material selection out of the available specifications with or without additional requirements is absolutely left to the technical judgment, knowledge, and experience of professional engineers. Some of the literature studies and consultants recommend material selection.

2.3.1 Criteria for Selection of Materials

Criteria for material selection are a very broad subject and cover several aspects and factors. No equation can be evolved for selection of material. There can be many material specifications available for given conditions with little difference. There can be overlapping. The designer has to compare their physical properties and apply his mechanical and metallurgical knowledge as well as cost and compare thickness required with each material to select particular specification. However, some of the main criteria are listed below:

1. Nature of fluid and chemical reaction
2. Service temperature
3. Strength of material
4. Formability, weldability, and machinability
5. Resistance to corrosion
6. Resistance to the outside environment
7. Safety and hazards of life
8. Service life

The following factors are useful for the selection of material.

2.3.2 Ductile and Brittle Materials

The material used in pressure-retaining parts needs to be ductile, in which marked plastic deformation commences at a fairly definite stress (yield point, yield strength, or possibly elastic limit), and exhibits considerable elongation (plastic deformation) before it fails reaching ultimate tensile strength (UTS). Observing this deformation, necessary steps can be taken to stop further deformation and possible accidents.

Ductile materials have a sufficient gap between yield (Sy) and UTS to observe the deformation. The ratio of Sy to UTS of normally used pressure parts is as low as 0.5 for SA-178A and SA-179 and max 0.7 for 91 grade LAS. The beginning of plastic deformation is not clearly defined in brittle metals, and the material exhibits little or no deformation before it fractures. Brittle materials are used in PPs with high temperature and low pressure whose thickness is designed mainly according to parameters other than pressure-like castings in valves.

2.3.3 Chemical Effect of Fluid in Contact

The chemical effect is mostly due to corrosion. LASs are used for hydrogen service, and austenitic SSs are selected for acidic fluids. Nonferrous materials are selected for various chemical fluids. For temperatures below 425°C, CS can be used with SS weld overlay or cladding or lining. Glass, rubber, lead, and Teflon can also be used as lining.

2.3.4 Temperature

In the cryogenic temperature range (−250°C to −100°C), carbon and LASs are brittle; therefore, austenitic SSs or nonferrous metals such as aluminum alloys, which do not lose impact strength at very low temperatures, are used. At low temperatures (−100°C to 0°C), LAS and fine grain CS that possess required impact properties are found to perform satisfactorily. In the range of intermediate temperatures (0°C to 425°C), low CSs can be used. At temperatures above 400°C, CSs exhibit a drop in yield and tensile strength and cease to be elastic and become partly plastic. Under a constant load, a continuous increase in permanent deformation called creep occurs. For high temperatures up to 650°C, LASs, molybdenum or chrome–molybdenum (up to 9% Cr+1%Mo), are used. For higher temperatures up to 816°C, austenitic SSs are used. For temperatures more than 816°C, nonferrous alloys, incoloy, or inconel, are used. However, codes permit up to temperatures 454 to 530°C for CS, up to 565°C for 11 and 22 grades of LAS, 600°C for 91 grade LAS, 830°C for SSs, and 900°C for high alloy steels.

For bolts and nuts, CS is not advised above 230°C (max up to 343) and for long periods of operation as relaxation will occur. Therefore, alloy steels are used. For higher temperatures over 500°C, austenitic SSs are used. Table 2.1 shows recommended American (ASME-SA) materials for different service temperature ranges.

2.3.5 Pressure and Allowable Stress

Allowable stress and ductility are generally opposite. The higher the allowable stress, the lower the ductility, and vice versa. Thickness depends on allowed stress for a medium and high pressure so a material with optimum higher allowable stress and cost can be selected. For very low pressure vessels, allowable stress is not effective, so a low strength material with low cost can be used. Alloy and SSs have comparatively lesser allowable stress at temperatures less than 400°C, than CS; hence they are not preferred normally.

TABLE 2.1
Recommended American (SA) Materials for Different Service Temperature ranges

Temp °C	Plate	Tube	Pipe	Forging	Bolt&nut	Strl Parts
−250 to −200	SA240 TP304, 304L, 347	SA213 TP304, 304L, 347	SA312 TP304, 304L, 348	SA182 F304, 304L, 349	SA320 B8 SH, SA194 gr8	Same as PPs
−200 to −100	SA240 TP304, 304L, 316, 316L, SA353	SA213 TP304, 304L, 316, 316L, SA333 gr.B	SA312 TP304, 304L, 316, 316L, SA333 gr.B	SA182 F304, 304L, 316, SA522	SH=strain-hardened, SA320B8 SH, SA194 gr8	Same as PPs
−100 to −45	SA203	SA-334 gr1	SA333 gr1	SA350 grLF1, 2	SA320 B8, SA194 gr4	Same as PPs
−45 to 0	SA515, 516 impact tested	SA210 A1, SA278, SA279	SA106B	SA105, 266	SA320 B8 SH, SA194 gr8	SA36 Si killed, fine grain for LT
0 to 425	SA515, 516, SA205, 302	SA210 A1, SA278, SA279	SA106 B	SA105, 267	SA193 B7, 194 gr2H	SA36
425–470	SA205, 302	SA213 T1	SA335 P1	SA182 gr F1	SA193 B7, 194 gr2	Same as PP
470–535	SA387 gr.11,22	SA213T11, 22	SA335P11, 22	SA182 F11, 22	SA190 B5, 194 gr3	Same as PP
535–650	SA387 gr.11, 22, 91	SA213 TP11, 22, 91	SA335 P11, 22, 91	SA182 F11, 22, 91	SA190 B5, 194 gr3	Same as PPs
600–816	SA240 304, 316, 321, 347	SA312 304H, 316H, 321H, 347H	SA312 304H,316H, 321H, 347H	SA182 304H, 316H, 321H, 347H	SA190 B8, 194 gr8	Same as PPs

Note: PP = pressure part.

2.4 HEAT TREATMENT

Pressure components fabricated from plate materials will undergo various manufacturing processes in steel mills and manufacturing units such as welding, rolling, and forming, during which internal stresses may be induced. During operation, these stresses together with membrane stress may exceed the yield point. Therefore, such fabricated parts, after rolling, bending, or forming and after the final stage but before the hydraulic test, are subjected to suitable heat treatment for relieving such internal stresses. Heat treatment is not required if the intensity of such internal stresses is marginal. The holding temperature and time of heat treatment depend on materials and are given in codes. The code[1] gives rules which are briefed in the following subsections.

2.4.1 Division of the Ferrous Material

For the purpose of heat treatment and radiography, the ASME code divides ferrous materials into part number (P) and group number (G). Some of the normally used materials are listed below:

a. CSs
   1. UTS ≤ 485 MPa (SA-192, SA-106B, SA-210 A1) P.No. 1 and G.No. 1
   2. UTS > 485 MPa (SA-105, SA-516 70, SA-266, SA-350) P.No. 1 and G.No. 2

b. LASs
   1. Gr-11 (1.25Cr, 0.5Mo) P.No. 4 and G.No. 1
   2. Gr-22 (2.25Cr, 1Mo) P.No. 5A and G.No. 1
   3. Gr-91 (9Cr, 1Mo) P.No. 15E and G.No. 1

c. Austenitic SS P.No. 8 and G.No. 1

d. Others
   1. low Mn, Cr, and Ni alloys (P1/G2, P1/G3, and P3)
   2. high Cr and Ni alloys (P6, 7, 9, 10, and 11)

2.4.2 Post-fabrication Heat Treatment

For CS, this requirement is mandatory in code when fiber elongation is > 40% [5% for certain conditions given in UCS-79(d)(2)] in rolling (bending) or forming at a temperature below 480°C. The equation for fiber elongation (εf) is given by

$$\varepsilon f = kt/R_f(1 - R_f/R_o),\ k = 50 \quad \text{for cyl. and} = 75 \text{ for head (double curvature)}$$
$$\varepsilon f = 50d/R \quad \text{for tube/pipe}$$

Where R = bend radius and R_o = radius of part before forming, infinity for plates, and d = OD of the pipe/tube, R_f = radius after forming, t = thickness

2.4.3 Post-weld Heat Treatment (PWHT)

PWHT is required for all welds of equipment with lethal gas in it, unfired steam boiler and material with part and group numbers (G3 of P3, G1 of P10B, G1 of P10F and P15E/G1) as per code. For other groups, PWHT is not required when certain parameters are within certain limits. The limits are higher for carbon steels and stringent as alloying is more.

For carbon steels of part & group numbers (P1/G1,2,3) as per table UCS-56-1, PWHT is not required under the following conditions: (Reprinted from ASME 2019 BPVC, Section VIII-division 1, by permission of The American Society of Mechanical Engineers)

a. For welds directly exposed to firing with thickness (T) ≤ 16 mm
b. For butt welds if [T ≤ 32 mm or {T ≤ 38 mm & pre-heated to (PH) > 95°C}]
c. For attachment welds with groove ≤ 13 mm & throat ≤ 13 mm, {for nozzle with (ID < 50 mm or PH > 95°C)}
d. For Tube to tube sheet (TS) welds if (tube OD ≤ 50 mm & TS Carbon ≤ 0.22% & PH > 95°C)
e. For non-pressure parts if (shell T ≤ 32 mm & PH > 95°C)

For other groups, similar but stringent rules are provided in Tables UCS-56. Boiler codes are more stringent than above code. Indian boiler code exempts only up to 20 mm thick.

2.5 NON-DESTRUCTIVE TESTING (NDT) AND WELD EFFICIENCY

NDT includes radiography (RT), ultrasonic (UT), die penetration (DT), magnetic particle (MPT), impact test, hydraulic test (HT), pneumatic test, etc., and extracts from the Tab UCS-57 & UW-12 of ASME 2019 BPVC, Section VIII-division 1, by permission of The American Society of Mechanical Engineers are given below.

For the purpose of NDT of welds and weld efficiency, welds are categorized as A = long. BW, B = circ. BW, C = flange weld, and D = branch weld. Abbreviations: BW = butt weld, AW = attachment weld, PH = preheating, CE = [C + (Mn + Si)/6 + (Cr + Mo + V)/5 + (Ni + Cu)/15] or [C + (Mn + Si)/6 + 0.15) ≤ 0.45, & individual pass ≤ 6]

a. RT of welds: UT can be performed in lieu of RT
   All BWs of vessels with the following parameters require 100% RT:
   1. Using lethal substances, having weld thickness (t) > 38(1.5"), [UFSB with internal pressure >50 psi(350kPa)], and [B and C category with > NPS 10" (DN 250) or t > 1.125"(29)]
   2. C&LAS material with part no. and thickness larger than as indicated below (refer: tab. UCS-57 of code): [(t > 32 mm for P. No 1), (>19 for 3,10A), (>16 for 4,9A,9B,10B.10C), and (> 0 for 5A,5B,5C,15E)]
   3. For boilers: (all longitudinal), (circumferential with DN > 250 or t > 29 mm), {if the fluid in the shell is steam and (DN > 400 or t > 41)}, and {if the weld is exposed to radiant heat and (DN > 100 or t > 13)}
b. Weld joint efficiency E:
   1. Double BW from both sides or equivalent (single welded with a backing strip that remains): E = 1 for 100% RT, E = 0.85 for 10% RT, and E = 0.7 for no RT

2. Single BW with a backing strip: E = 0.9 for 100% RT, E = 0.8 for 10% RT, and E = 0.65 for no RT
3. Single circ. BW without a backing strip with t ≤ 16 & OD ≤24" (600) and E=0.6 for no RT

2.6 IMPACT TEST AND MINIMUM DESIGN METAL TEMPERATURE (MDMT)

a. *Fracture toughness* is the ability of a material to withstand fracture in the presence of cracks. With a decrease in temperature, strength of material increases but fracture toughness decreases and experiences a shift from ductile to brittle. It is sometimes defined as the temperature at which the material absorbs 15 *ft. lb* of impact energy during fracture. However, physical properties and microstructure will remain the same after returning to normal temperature. Flaws may appear as cracks, voids, metallurgical inclusions, weld defects, design discontinuities, or some combination thereof. Many engineering metals and alloys that become brittle shatter unexpectedly at low temperatures when loaded to stress levels below allowable at normal temperatures. Addition of the Ni alloy improves toughness. Toughness is measured by the impact test in ft.lb (Nm).

b. *Impact test*: Generally, raw materials for PPs are selected as specified in codes whose specification such as SA516 70 specifies all the requirements of testing suitable for PPs including impact testing for low temperature use and those carried out in steel mills. For the material not tested for, the impact property can be used for duly testing if required by the code. Figure UCS-66 of code gives the minimum required V-notch test value for the component thickness and its UTS.

c. *MDMT* (T_d): All parts of pressure vessels are to be designed not only for maximum operating conditions (pressure and temperature) but for all transient conditions such as start-up, shutdown, idle, etc. The lowest temperature among all the conditions is called design MDMT (T_d). Parts of certain material designed to the rules of code may fail when its metal temperature is lowered to certain limit under coincident pressure due to inadequate toughness. This temperature coincident with pressure is called MDMT (T_a) of that part. If $T_a > T_d$, the impact test is to be carried out. The value of T_a depends on the type of material, temperature, stress ratio (induced membrane stress to allowed stress), thickness, and other properties. The components shall meet toughness to a minimum limit at design MDMT so that its ductility will not shift to brittle and shall be checked before operating them. The code gives rules in UCS-66

d. *Calculation of MDMT of component* T_a:

1. Material: The code categorizes the raw material into four groups (A, B, C, and D) depending on their metallurgical composition, heat treatment, and product form, and min temperature limit (T_{ma}) can be read from Figure UCS-66 or table UCS-66 for A, B, C, and D against governing thickness (t_g) of the component, below which they are prone to the above brittle failure. Some random values from table UCS-66 of ASME 2019 BPVC, Section

TABLE 2.2
Values of T_{ma} °C

Tg"	A	B	C	D
0-3/8	−8	−29	−48	−48
0.5	0	−22	−37	−48
1	20	1	−18	−33
2	37	17	−3	−20
3	43	26	5	−12
4	47	32	11	−5
5	49	37	16	1
6	49	40	19	4

VIII-division 1, by permission of The American Society of Mechanical Engineers are listed in Table 2.2.

2. Calculation of governing thickness t_g: for welded parts = t nominal thickness and for nonwelded parts $t_g = t/4$
3. Stress ratio: The stress ratio will influence T_a. The lower the stress ratio, the lower the limit and for the stress ratio less than 0.35, brittle failure may not occur. For stress ratio > 0.35 and < 1, T_{ma} can be reduced by Tr which reduces with the increase of stress ratio. T_a is calculated by deducting Tr from T_{ma} and given by $T_a = T_{ma} - Tr$

Tr against the stress ratio can be read in Figure UCS-66.1 of ASME 2019 BPVC, Section VIII-division 1, by permission of The American Society of Mechanical Engineers and given below

$$Tr = \{140, 95, 58, 40, 30, 20, 10, 0\}°F \text{ for stress ratio } \{0.35, 0.4, 0.5, 0.6, 0.7, 0.8, 0.9, 1\}$$
$$Tr = \{80, 52, 32, 23, 17, 6, 0\}°C \text{ for stress ratio } \{0.35, 0.4, 0.5, 0.6, 0.7, 0.8, 0.9, 1\}$$

If T_a for any part > T_d of equipment, the material of that part is to be impact tested as per UG-84 and shall meet the result given in Figure UG-84.1M.

(e) Before calculating T_{ma}, if the raw material of part is satisfactorily impact tested as per its specification or the part has certain properties and conditions given in UCS-66, it is concluded that the part has the required toughness and the impact test is not mandatory. Such properties and conditions are given below.

1. Nonwelded parts with $t_g < 150$ mm and $T_d > 50$°C [ref: (a)(1)–a(e)]
2. welded parts with $t_g < 100$ mm and $T_d > 50$°C [ref: (a)(1)–(a)(5)]
3. $t_g \leq 2.5$ and $T_d \geq -48$ [ref: (d)(1)–(e)]
4. $-105 < T_d < -48$ & stress ratio ≤ 0.35, [ref: (b)(3)]
5. plate t $\leq$ 2.5 and temp $\geq$ −48 (tube & pipe ref rule) not required [ref: UCS-16(d)]

(f) Impact test required in the following conditions
1. yield tress >450 MPa required, [ref: (f)]
2. $T_d < -105$ impact test required, [ref: (b)(3)]
3. $T_d < -48$ and stress ratio >0.35, [ref: (b)(2)], however may be impact tested if $T_d < -48$ even if the component is not under pressure.

If none of the factors in (e) and (f) is applicable to the part, T_{ma} of that part is calculated to verify whether the impact test is required or not.

2.7 HYDRAULIC TEST (HT)

The HT is carried out after completing manufacturing and before dispatching in the presence of the inspecting agency to verify any leakage. Apart from leakage, radial growth (yield) of the shell is measured by noting the circumference of the shell before, during and after the hydraulic test. Any leakage other than through flange joints and yielding indicates error in any one or more activities right from raw material manufacturing until this stage of HT. The HT of any equipment is NDT. Therefore, stress in any component or at any point of equipment under HT pressure (P_h) shall not exceed its yield stress (Y) theoretically, practically limited to 0.9Y.

Let P_h = K x MAWP, P = MAWP = max allowed working pressure of equipment

There are three methods of computing constant K based on the concept explained above.

1. *Constant method*: Because the factor of safety is normally 1.5, K is taken as 1.5 in most of the codes. The maximum stress at any point in this method may exceed 0.9Y in the case of almost the same temperature at HT (T_h) and working condition (T), or the gap between max stress and Y is higher with higher temperature difference leaving scope for increasing HT pressure for more safety. Some codes such as ASME S I provide this rule to limit stress to 0.9Yat P_h.
2. Variable method: K is not constant and equal to 1.3 times least stress ratio (LSR). 1.3 in place of 1.5 is to limit max stress to 0.9Y.

   LSR = S_h/S, where S_h = allowed stress at test temp. T_h and S = allowed stress at design temp. T.

   Because the LSR value is 1 for $T_h = T$ and also for some high strength materials having S is the same up to certain temperature. As the difference in S_h and S increases, the LSR increases and max stress in HT is nearer to 0.9Y independent of the T value. ASME S VIII D 1 uses this rule.
3. Least calculated pressure method: P_h = 1.3 times least calculated pressure (LCP -least MAWP at any point or component). when margin is provided by designer second method results in lesser stress at HT. P_h is increased in proportion to margin to increase stress at HT nearer to 0.9Y. Hence, LCP is used instead of LSR times MAWP. ASME S VIII D 1 uses this method as an option.

REFERENCE

1. Code ASME S VIII D 1, 2019

3 Mechanical Design Basics

3.1 BASICS

Readers are advised to review basics of direct (tensile and compressive), shear, bending, and torsion stresses and strains; force, moment, and torque; yield and tensile strength, elastic modulus, rigidity modulus, Poisson's ratio, first and second moment of area; and polar second moment of area before reading this book. However, some of the above and other basics difficult to feel are described below to get a practical feeling.

3.1.1 XYZ Coordinate Convention

One of the three conventions normally used is: forces and displacements x and z are horizontal and y vertical + up. When looking from top (+y) $+z$ rotated by 90° in the anticlockwise will coincide with $+x$ (yzx). Similar is the convention for zxy and xyz systems. Planes are referred as xy (z-axis), yz (x-axis) & zx (y-axis). Moments and rotations by the right hand thumb rule: thumb pointing in + direction of force (F_X) and rotation of fingers is + for moment (M_X).

3.1.2 Degrees of Freedom

Degrees of freedom of a boundary point (node point) are six, three-translational x, y, and z; and three-rotational θ_X, θ_Y, and θ_Z for space elements; and three, two-transitional x, y and one-rotational θ_Z for the xy plane (Z axis) element.

3.1.3 Boundary Conditions

Boundary Conditions of Each End of Element Are Fixed, Pinned, Simply Supported, Guided, and Free

- *Fixed*: all the six degrees of freedom are zero
- *Pinned*: translations are zero and rotations are free
- *Simply supported*: all rotations and all translations except vertical are free
- *Free*: all the six degrees of freedom are free
- *Guided*: all rotations are zero and axial translation is zero

Boundary conditions of a node can be in any combination. For the beam element to be stable and determinate, the sum of the degrees of freedom of both ends shall be six.

Elements with three degrees of freedom are called statically determinate, and reaction forces and internal forces can be found out from static equilibrium equations alone. Less is in-determinate which requires deformation equations in addition to

DOI: 10.1201/9781003091806-3

static equilibrium to solve for unknown forces. Fixed beam is in-determinate and analyzed by deriving the deflection equation by the integration method. Beams with three supports (long horizontal vessel) are in-determinate and analyzed using three-moment (Clapeyron's) theorem. See Example 3.4. The rest is unstable, cannot be analyzed, and needs further restraints.

Boundary conditions can also be restricted (like $x = +3$ mm mean the point can move 3 mm in positive x direction and further movement is restricted), and each degree of freedom can be plus or minus or both.

3.1.4 Load Condition

Like boundary conditions, loads are three forces F_X, F_Y, and F_Z and three moments M_X, M_Y, and M_Z. *Forces* include moments and are called force tensor; and *displacements* include rotations and are called displacement tensor.

$$\text{Force tensor (matrix)}[F] = \{F_X, F_Y, F_Z, M_X, M_Y, M_Z\}$$

$$\text{Displacement tensor (matrix)}[D] = \{\theta_X, \theta_Y, \theta_Z, x, y, z\}$$

3.1.5 Transfer of Forces

Transfer of forces from 1 to 2 is given by equations below:

$$(F_X, F_Y, F_Z)_2 = (F_X, F_Y, F_Z)_1$$

$$M_{X2} = M_{X1} + F_{Y1}\,z + F_{Z1}\,y$$

$$M_{Y2} = M_{Y1} + F_{Z1}\,x + F_{X1}\,z$$

where $x = x_2 - x_1$, $y = y_2 - y_1$, and $z = z_2 - z_1$

$$\text{Matrix equation for } [F_2]_{61} = [T]_{66}[F_1]_{61}$$

Transformation matrix [T] is given in Table 3.1

TABLE 3.1
Matrix for T & R, C = cos, and S = sin

	Trans. matrix-T						Rotation matrix-R					
	1	2	3	4	5	6	1	2	3	4	5	6
1	1	0	0	0	0	0	Cθ	Sθ	0	0	0	0
2	0	1	0	0	0	0	–Sθ	Cθ	0	0	0	0
3	0	0	1	0	0	0	0	0	1	0	0	0
4	0	Z	y	1	0	0	0	0	0	Cθ	Sθ	0
5	z	0	x	0	1	0	0	0	0	–Sθ	Cθ	0
6	y	X	0	0	0	1	0	0	0	0	0	1

Note that the [R]33 for forces and moments is same.

3.1.6 Rotation of Axis

a) One plane is the same say XY(Z axis) or rotation of axis in one plane (XY):
   Rotation of the corresponding (Z) axis by angle θ, from XYZ to X'Y'Z'
   Angle anticlockwise from x to x' or y to y' = θ
   Angle anticlockwise from y to x' = 270 + θ
   Angle anticlockwise from x to y' = 90 + θ

The rotation vector can be expressed in matrix form

$$\begin{matrix} \cos\theta & \sin\theta & 0 \\ -\sin\theta & \cos\theta & 0 \\ 0 & 0 & 1 \end{matrix}$$

Similarly, the matrix for Y and X axis can be derived as

$$\begin{matrix} \cos\theta & 0 & -\sin\theta \\ 0 & 1 & 0 \\ \sin\theta & 0 & \cos\theta \end{matrix} \quad \text{and} \quad \begin{matrix} 1 & 0 & 0 \\ 0 & \cos\theta & \sin\theta \\ 0 & -\sin\theta & \cos\theta \end{matrix}$$

To convert any vector [F] from global to local [F'] or in reverse, this matrix can be used in equation [F'] = [R] [F].

It may be noted that $[R]_{33}$ for forces and moments is the same.

Example 3.1: Calculate force tensor $F_X = F_Y = F_Z = M_X = M_Y = M_Z = 1$ for rotation of *Z* axis by angle θ = 30° = 0.5236 radians.

$$F_X' = F_X \cos\theta + F_Y \cos(270+\theta) = F_X \cos\theta + F_Y \sin\theta = 1.366$$
$$F_Y' = F_Y \cos\theta + F_X \cos(90+\theta) = F_Y \cos\theta - F_X \sin\theta = 0.36$$
$$F_Z' = F_Z = 1$$
$$M_X' = M_X \cos\theta + M_Y \sin\theta = 1.366$$
$$M_Y' = M_Y \cos\theta - M_X \sin\theta = 0.366$$
$$M_Z' = M_Z = 1$$

Matrix equation for $[F']_{61} = [R]_{66}[F]_{61}$
Rotation matrix [R] is given in Table 3.1

b) None of the planes are the same
   1) *Rotation of axis in two planes* for converting vector OA (L) along +x axis in the global system (XYZ) to vector in the (ABC) system so that it aligns with +A axis. Vector direction cosines in the (ABC) system are: x/L, y/L, z/L, where x, y, and z are unit vectors in the XYZ system.

To derive the [R] matrix for rotation of Force vector [F] to F'], rotate Z axis by angle (θ) to align with force F' component in the XY plane. Where cos θ = x/L_{XY},

$L_{XY} = \sqrt{(x^2 + y^2)}$. Then, rotate the new Y axis (formed by first rotation) by angle (ρ) to align with force F'. where $\cos\rho = Lxy/L$, $L = \sqrt{(x^2 + y^2 + z^2)}$. The rotation matrix for both rotations $[R_1]$ and $[R_2]$ is as derived above in (a)

$$\begin{matrix} \cos\rho & 0 & -\sin\rho \\ 0 & 1 & 0 \\ \sin\rho & 0 & \cos\rho \end{matrix} \quad \text{and} \quad \begin{matrix} \cos\theta & \sin\theta & 0 \\ -\sin\theta & \cos\theta & 0 \\ 0 & 0 & 1 \end{matrix}$$

$$[R_1][R_2] = \begin{matrix} \cos\theta\cos\rho & \cos\rho\sin\theta & -\sin\rho \\ -\sin\theta & \cos\theta & 0 \\ \cos\theta\sin\rho & \sin\theta\sin\rho & \cos\rho \end{matrix}$$

The force matrix equation is

$$[F'] = [R_1][R_2][F]$$

Example 3.2: Calculate F′ at point P given unit force ($F_x = F_y = F_z = 1$) at O (0,0,0).

Forces at point of P′ projection of P on the XY plane after first rotation $\theta = 30°$ are the same as Example 3.1.

That is, $F_{X1} = 1.366$, $F_{Y1} = 0.36$, and $F_{Z1} = 1$

For the computation of forces at point of P for second rotation of 30°, the same procedure is used (rotation is in new Y axis).

$$F_x' = F_{X1}\cos\rho - F_{Z1}\sin\rho = 0.683 = F_x\cos\theta\cos\rho + F_y\sin\theta\cos\rho - F_z\sin\rho$$
$$F_y' = F_{Y1} = 0.183 = Fy\cos\theta - Fx\sin\theta$$
$$F_Z' = F_{Z1}\cos\rho + F_{X1}\sin\rho = 1.55 = Fz\cos\rho + Fx\cos\theta\sin\rho + Fy\sin\theta\sin\rho$$

It may be observed that the equations in Example 3.2 by basics are the same as matrix equations given above.

2) *General rotation of the 3d system*: Unlike rotation of axes in (1), analysis involves the rotation of planes. OA is a straight line in space of global coordinate system XYZ with origin O and coordinates of A (x, y, z) and ABC local system with OA as + x-axis. The rotation matrix $[R]_{33}$ can be computed as per the Eularian angle method, by rotating first Z axis by angle α, then rotating new Y axis by angle β, and third new Z axis by angle γ. A three rotation matrix $R_Z(\alpha)$. $R_Y(\beta)$. $R_Z(\gamma)$ can be computed as given in (a) and is multiplied to obtain a [R] general matrix and given by

$$R = R_Z(\gamma)\cdot R_Y(\beta)\cdot R_Z(\alpha) = [(C\alpha\, C\beta\, C\gamma - S\alpha\, S\gamma), (-C\alpha\, C\beta\, S\gamma - S\alpha\, C\gamma), (C\alpha\, S\beta)]$$
$$[(S\alpha\, C\beta\, C\gamma + C\alpha\, S\gamma), (-S\alpha\, C\beta\, S\gamma + C\alpha\, C\gamma), (S\alpha\, S\beta)]$$
$$[(-S\beta\, C\gamma),\quad (S\beta\, S\gamma),\quad C\beta]$$

where α, β, and γ = angles of rotations in rotating XYZ to coincide with ABC. C = cos and S = sin.

3.1.7 Free Body Diagram

The free body diagram shown in Figure 3.4 of an element of a structure (equipment) is the diagram of only that element, as if made free from the rest of the structure, with all the internal and external forces acting on it including end restraints. It is the same as the loading diagram with boundary supports replaced by reactions. Apart from analysis under forces, it is used for solving unknowns. For the plane element, three equations can be compiled by static equilibrium:

1) and 2) Vector sum of forces in both directions is zero and 3) moment of all forces (include moment) at any point is zero.

Several equations can be formed by taking moment at several points. It is true, but cannot combine them to obtain the fourth unknown. For the fourth unknown, use other criteria such as deformation. For solid elements, six equations can be formed.

3.1.8 Poisson's Ratio

Poisson's ratio can be explained by the example of a rectangular bar (10 × 50 × 1000 mm) with E = 20000 kg/mm² subjected to tensile load 5000 kg along length which will increase by 0.5 mm. At the same time, its section will reduce to 9.9985 × 49.9925. Sides 10 and 50 each are reduced by [0.3(0.5/1000)10 = 0.0015] and 0.0075 mm, respectively. This 0.3 in the above equation is Poisson's ratio. It is the ratio of transverse strain to axial strain. For small values (theoretically for cuboids), it is the transverse movement by axial movement. Volume 10*50*1000 = 500000 reduces to 9.9085*49.9925*1000.5 = 495598. If the load is compressive, the volume will increase with the corresponding change in its density for conservation of weight. Symbol (v) is used for Poisson's ratio in this book.

3.1.9 Moment of Inertia

Moment of inertia (MI or I) or second moment of area of a plane section at its neutral axis (NA) is integration of second moment of area of section at *NA*. In dynamics, this is the mass of body instead of area of plane section. Thus, *MI* is the property of section resisting bending moment and deflection.

For rectangle *B X L*, *MI* at its *CG* (middle axis) by definition is integration of the second moment of area (B dx) at a distance x from middle is given by

$$I = \int_{-L/2}^{L/2} Bx^2 dx = B(x^3/3 + C)$$

Because $I = 0$ when $x = 0$, constant of integration $C = 0$, therefore

$$I = B[x^3]/3 = B[(L/2)^3 - (-L/2)^3]/3 = B\,L[L^3/8 + L^3/8]/3 = B\,L^3/12$$

Similarly for any section, it can be derived.

$$I = \pi D^4/64 \text{ for circle of diameter } D$$

Integration is not required for all sections. Use logic and maths as $\pi(D_o^4 - D_i^4)/64$ for a cylindrical shell with outside diameter (D_o) and inside diameter (D_i) using MI of solid cylinders.

In general to calculate MI at any axis of any section, divide the section into elements with available formula for *I* at its NA parallel to the required axis, or with NA in the required direction passes through its CG, or symmetrical in the required direction. Calculate NA of section by the first moment method and then MI of elements at NA of section by the equation below.

I_n = I of element at its NA (I_C) + area of element multiplied by square of the distance of its NA to NA of section. Add I_n of all elements to obtain *I* of section at its NA.

Example 3.3: Calculate MI of angle 100 × 100 × 10 mm as shown in Figure 3.1 and divide into two rectangles 100 × 10 and 10 × 90 with I_C = 8333 and 607500. The distance of *NA* from the flange bottom is found out by taking moment at the flange bottom.

$$y = \frac{100*10*5+90*10*(10+45)}{(100*10+90*10)} = 28.68$$

I of element 1 = 8333 + 100*10*(28.68–5)^2 = 569265
I of element 2 = 607500 + 90*10*(55–28.68)^2 = 1230779
Adding both = *I* of section = 569265 + 1230779 = 1800044
Being equal angle, *I* is the same in either axes. $I_{xx} = I_{yy}$

3.1.10 Polar Moment of Inertia

Polar moment of inertia (J) of a section used to resist torsion moment (T) at that section is equal to the sum of *MI* of the section in both perpendicular axes and given by

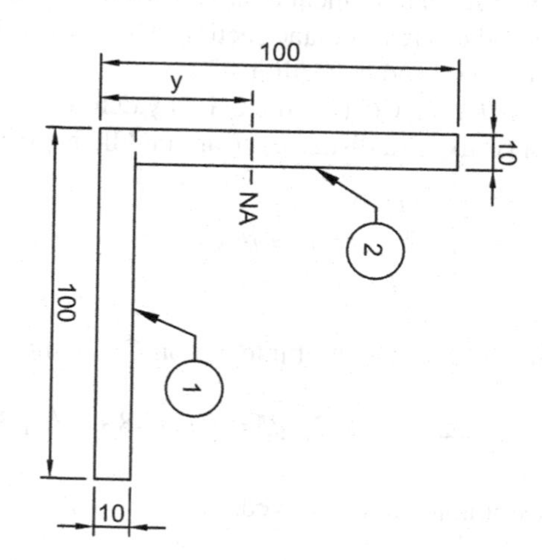

FIGURE 3.1 Angle for MI.

$J = I_{XX} + I_{YY}$. Its value for angle shown in Figure 3.1 with dimensions of Example 3.3 being equal to 2 × 1800044 = 3600088.

3.1.11 Significant Figures in Numbers

No calculated value of stress, strength, or deformation can be regarded as exact. The formulas used are based on certain assumptions as for properties of materials, regularity of form, and boundary conditions that are only approximately true. They are derived by mathematical procedures that often involve further approximations. In general, therefore, great precision in numerical work is not justified. Three significant figures concerning the use of formulas may be of value like 34500, 5.49, and 0.00568.

3.2 DESIGN THEORIES

General: Elastic and Plastic theory

Particular: Beam, Membrane (Chapter 4), and Flat Plate Theory (Chapter 10)

The subject of analysis of any problem using the above design theories is generally called structural engineering, and the analysis is called structural analysis.

3.2.1 Elasticity Theory and Hook's Law

Hook's law states that within the elastic limit stress is proportional to strain. It holds for the material of which a body is composed, the body will usually confirm to a similar law of load-deformation proportionality, and in most instances, the deformation is proportional to the magnitude of the applied load or loads.

There are two important exceptions to this rule. One is to be found in any case where the stresses due to loading are appreciably affected by the deformation like beam subjected to axial and transverse loads and a flexible wire or cable held at the ends loaded transversely. The second exception is represented by any case in which failure occurs though elastic instability, as in a slender (*Euler*) column or in buckling of the cylindrical shell under external pressure or local load.

External forces with points of application that do not move are called reactions, and external forces with points of application that move are called loads.

When an elastic system is subjected to static loading, the external work done by the loads as they increase from zero to maximum value is equal to the stain energy acquired by the system. For single load, the deflection at the point of loading in the direction of load is equal to twice the strain energy divided by the load.

When an elastic system is statically loaded, the partial derivative of the strain energy with respect to any one of the applied forces is equal to the movement of the point of application of the force in the direction of that force.

3.2.2 Plastic Theory

In elastic theory, maximum stress occurs at extreme fiber for bending moment and allows up to yield stress divided by the factor of safety. Ductile materials such as

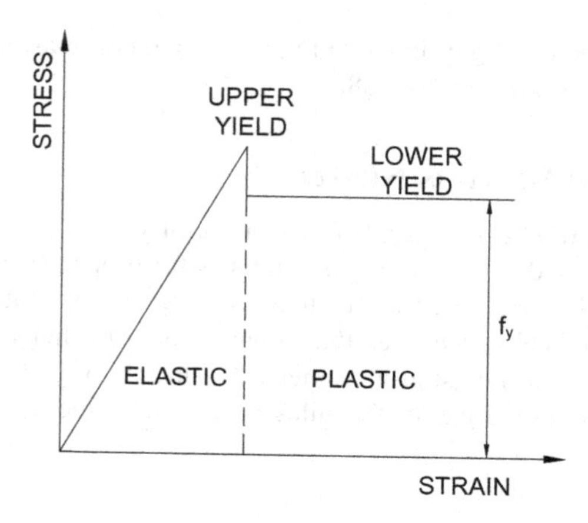

FIGURE 3.2 Stress strain diagram in plastic theory.

mild steel components and extreme fiber will not fail even after reaching yield stress, but will continue to withstand until the central section remains within the elastic limit.

As the load increased gradually, the outer fiber of the section will reach the upper yield point and then reduced to lower yield as shown in Figure 3.2. This outer fiber is in plastic state and shows a considerable increase in strain and deflection at that section with redistribution of stress. In mild steel, this increase in strain takes place without the increase in stress as shown in the figure. When the whole cross-section becomes plastic, no further increase in resistance is possible and a plastic hinge is formed. One or more such hinges are required for a complete collapse. The number depends upon the type of structure. The load at which this state collapses is called the *collapse load*, and the ratio of this load to the working load is called the *load factor*. In plastic design, this factor is used instead of the normal factor of safety.

Figure 3.3 shows the variations in stress and strain in a beam of symmetrical cross-section subjected to a bending load. As per theory of bending, the maximum working stress is f_t & f_C as shown in Figure 3.3a, and elastic theory is applicable. As load is increased, the extreme fibers reach lower yield stress f_Y, and the beam is in a partial plastic state as shown in Figure 3.3b and then complete plastic state with stress f_Y uniform over the whole cross-section shown in Figure 3.3c. The moment of resistance and shape factors in plastic theory are explained below.

(a) *Max moment of resistance* (M) of any cross-section as per elastic theory is equal to allowed stress multiplied by section modulus $= f Z$.

For rectangle of size b × d, $M = f\, b\, d^2/6$

Moment can also be derived from the stress diagram from elastic theory as follows.

When stress at edges is f_t tension and f_C compression, stress at any depth is reducing from edge to *NA* and zero at *NA* as shown in Figure 3.3a.

Consider d_X depth at x from *NA*

b = width at x from *NA* and $b = f(x)$, for simplicity consider b as a constant.

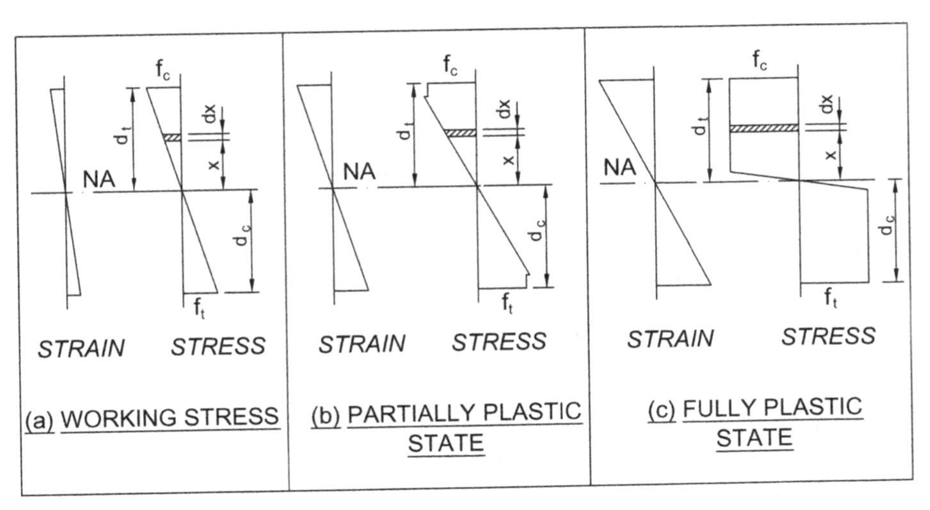

FIGURE 3.3 Variations in stress and strain diagrams in a beam.

d_t or d_c depths both sides from NA
Stress at x, $f_X = f_t\, x/d_t$ and $f_C\, x/d_C$
Resisting force $F = f_X\, b\, dx$
Moment due to F at NA = F x = $f_x\, b\, x\, dx$

$$M = b\int_{d_t}^{0}(f_t/d_t)x^2dx + b\int_{d_c}^{0}(f_c/d_c)x^2dx$$
$$M = b\left[(f_t/d_t)d_t^2/3 + (f_c/d_c)d_C^3/3\right] \tag{3.1}$$
$$M = \frac{b}{3}\left[f_t\, d_t^2 + f_C\, d_c^2\right]$$

For rectangle $d_t = d_C$ and $d = 2d_C = 2d_t$. Then, $f_t = f_C = f_Y$

$$M = \frac{b}{3}\left(2f_Y\frac{d^2}{4}\right) = f_Y\frac{b\,d^2}{6} = f_Y\, Z \tag{3.2}$$

M can also be derived and is equal to the moment of area of stress diagram at NA multiplied by width.

A = Area of stress diagram Figure 3.2a = $2\left(\frac{f_Y}{2}\frac{d}{2}\right)$

$$M = b\,A\frac{2}{3}\left(\frac{d}{2}\right) = f_Y\, b\, d^2/6 \tag{3.3}$$

Note that Eqs. 3.2 and 3.3 for the rectangle are the same.

(b) *Max moment of resistance* (M_P) of any cross-section as per plastic theory for the above can be derived as the moment of area of stress diagram Figure 3.3c at *NA* multiplied by width.

$$M_P = 2b\left(f_Y \frac{d}{2}\right)\left(\frac{d}{4}\right) = f_Y\, b\, d^2/4 = f_Y\, Z_P \tag{3.4}$$

(c) *Shape factor S_f*: Shape factor is the ratio of M_P/M or Z_P/Z. Therefore, for a rectangle, S_f can be obtained from Eqs. 3.2, 3.3, and 3.4 and is given as Eq. 3.5

$$S_f = \frac{M_P}{M} = \frac{Z_P}{Z} = \left(f_Y\, b\, d^2/4\right)/\left(f_Y\, b\, d^2/6\right) = 1.5 \tag{3.5}$$

The shape factor for other sections can be derived as above.

Shape factor for circular section = $16/3\pi$

The shape factor for the cylindrical shell with inside and outside radius r_i and r_O can be derived and is equal to Eq. 3.6

$$S_f = \frac{16 r_O}{3\pi} \frac{r_O^{\,3} - r_i^{\,3}}{r_O^{\,4} - r_i^{\,4}} \tag{3.6}$$

For 2 m dia × 10 mm, the shell shape factor by solving Eq. 3.6 = 1.3

Allowable bending stresses are therefore higher than that for tensile stress and are equal to or less than the shape factor.

Plastic theory is used for bending stresses.

3.2.3 Beam Theory

Any member of equipment subjected to bending in the axial plane under the action of lateral force is called beam.

The formulas of beam theory are based on the following assumptions:

- The elasticity and rigidity modulus and Poisson's ratio are the same in tension and compression and in all directions.
- The beam is straight or nearly so; if it is slightly curved, the curvature is in the plane of bending and the radius of curvature is at least 10 times the depth.
- The cross-section is uniform.
- The beam has at least one longitudinal plane of symmetry.
- All loads and reactions are perpendicular to the axis of the beam and lie in the same plane, which is a longitudinal plane of symmetry.
- The beam is long in proportion to its depth; the span to depth ratio is at least eight for compact sections and 15 for sections with thin webs.
- The beam is not disproportionately wide.
- The maximum stress does not exceed the proportional limit.
- Restraint is not on a point; it is plane with area.
- Concentrated load is not on a point, but on a plane.

- The sectional shape is not rigid. Local displacements (distortion) due to loading induce local stresses which reduce as the distance increases.

Applied to any case for which these assumptions are not valid, the formulas given yield results that at best are approximate.

3.3 FLEXURE OR BEAM FORMULAS

Consider the free body diagram of beam element dx in any beam with UDL = w as shown in Figure 3.4

By equilibrium of vertical forces

$$F + dF = F + w\,dx, \text{ or } dF/dx = w \tag{3.7}$$

By taking moments at the left edge of element for all loads

$$M + dM - (F + dF)dx - M + w\,dx\,dx/2 = 0$$

Neglecting the product of two small quantities

$$\frac{dM}{dx} = F \tag{3.8}$$

Consider small fiber AB with area dA at a distance y from the NA with radius R at NA of the bending curve as shown in Figure 3.5.

Strain in AB is given by Eq. 3.9

$$\frac{(R + y)d\theta - R\,d\theta}{R\,d\theta} = \frac{y}{R} = \frac{\sigma}{E} \tag{3.9}$$

By equilibrium of moment load and resisting moment

$$M = \int_{-y}^{y} \sigma\, y dA$$

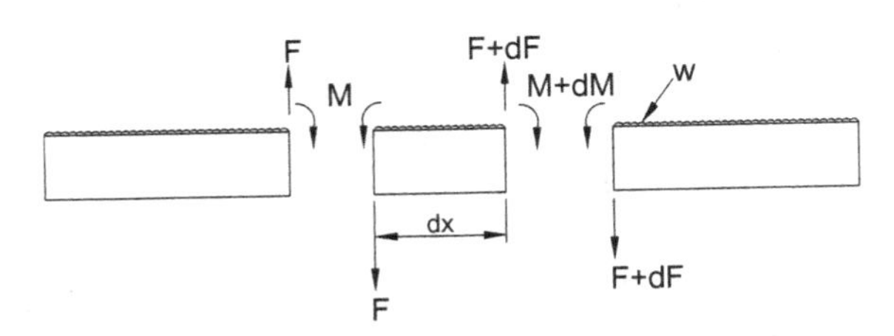

FIGURE 3.4 Free body diagram of the beam element.

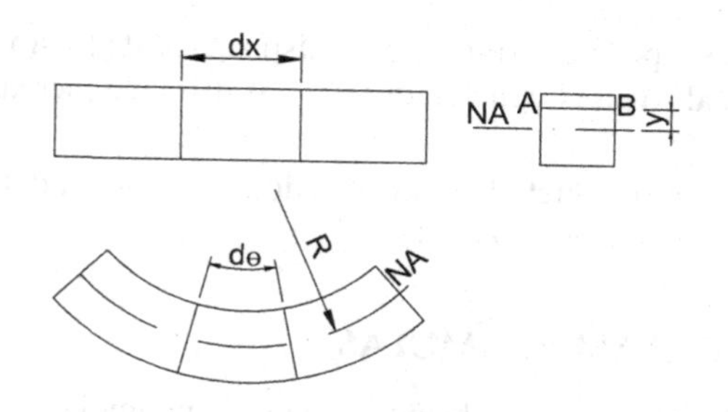

FIGURE 3.5 Bending curve of the beam

Substituting $y\ E/R$ for σ from Eq. 3.9

$$M = \int_{-y}^{y} \frac{E}{R} y^2 dA = \frac{E}{R} \int_{-y}^{y} y^2 dA \tag{3.10}$$

Substituting I moment of inertia of section containing AB for $\int(y^2 dA)$ in Eq. 3.10

M = E I/R or M/I = E/R, and from Eq. 3.9

$$\frac{\sigma}{y} = \frac{M}{I} = \frac{E}{R} \tag{3.11}$$

Eq. 3.11 is called the flexure formula

3.4 DEFLECTION

Deflection can be derived by three methods. Two methods are described below and third Castigliano's first theorem is covered in Chapter 9

3.4.1 Macaulay's Method of Integration

Consider the deflected beam under any load. Choose any convenient point as origin. dx and dy are incremental values at any point at x and y on the deflected curve in Figure 3.5. The deflected curve is considered as circular with radius R. The geometrical relation with R, dx, and dy is given by Eq. 3.12

$$\frac{1}{R} = \frac{d^2y/dx^2}{[1+(dy/dx)^2]^{1.5}} \tag{3.12}$$

Within the elastic limit, the slope dy/dx is very small and $(dy/dx)^2$ can be neglected in comparison, and Eq. 3.12 can be simplified to

$$1/R = d^2y/dx^2 \tag{3.13}$$

Substituting R = EI/M from Eq. 3.11 in Eq. 3.13, we obtain

$$EI\, d^2y/dx^2 = M \quad (3.14)$$

By successive differentiation, we get shear force and UDL given by

$$EI\, d^3y/dx^3 = dM/dx = F \quad (3.15)$$

$$EI\, d^4y/dx^4 = dF/dx = w \quad (3.16)$$

By double integrating Eq. 3.14, y can be solved

3.4.2 Moment Area Method

The loading, shear force, and moment diagrams of simply supported beam are shown in Figure 3.6

Rate of change of F or slope of F-diagram = rate of loading at any point.

At C, dF/dx from A to C = $(w\, L/2–0))/(L/2)$ = w

Rate of change of M or slope of M-diagram = F at any point.

At C, dM/dx = slope of M-diagram or F at C = 0

Change of slope (i) of beam between any two points = [area of the M-diagram between the two points]/EI

At C, $EI(i_c–i_a)$ = (area of the M-diagram from A to C) = $(w\, L^2/8)(L/2)(2/3) = w\, L^3/24$

Deflection $y_2–y_1$ = (Moment at origin of the M-diagram area between any two points 1 and 2)/$EI = (x_2\, i_2–y_2) – (x_1\, i_1–y_1)$

Taking origin at A, points A and C, $(y_C - y_a) = (x_C\, i_C - y_C) - (x_a\, i_a - y_a) = (w\, L^3/24)(5L/16) = 5w\, L^4/384$

Because $i_C = 0, x_a = 0, y_a = 0, (x_C\, i_C - y_C) - (x_a\, i_a - y_a) = (0 - y_C) - (0 - 0) = y_C$

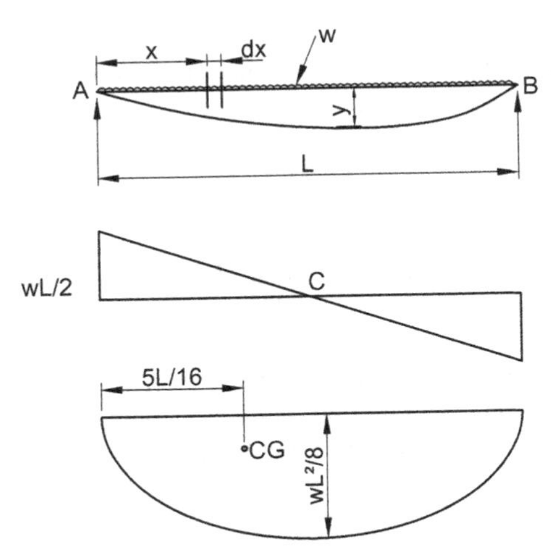

FIGURE 3.6 Loading, shear, and moment diagram.

TABLE 3.2
Value of K in Beam Formulas

Load Type	Conc. load W			Uniformly dist. load w		
Beam Type	**C**	**S**	**F**	**C**	**S**	**F**
$M_S = KWL$	1	0	1/8	1/2	0	1/12
$M_C = KWL$	0	1/4	1/8	0	1/8	1/24
$y = KWL^3/(EI)$	1/3	1/48	1/192	1/8	5/384	1/384
$\theta_e = KWL^2/EI$	1/2	1/16	0	1/6	1/24	0

M – moment, y – deflection, and θ – slope
L, span; W, w L for UDL; C, cantilever; S, simply supported; F, fixed; y, max deflection; subscript: s at support, c at centre, e at end, W at c for S & F and end for C

From the above two equations

$$y_C = \frac{5}{384} w L^4 \tag{3.17}$$

Formulas for moment, slope, and deflection for beams derived by the above theory are tabulated for reference in Table 3.2.

3.4.3 Strain Energy Method

When any load applied within elastic limit on an element, it deforms and returns back on releasing the load. The work done (energy input) is stored in the body as strain energy (U). Let a bar of cross section area (A), length (L), is under tension load (F), elongates by d and tensile stress (σ) induced.

Work done = $U = F\, d/2 = (A\,\sigma)(\sigma\, L/E)/2 = (\sigma^2/2E)$(volume)

Deformation $d = 2U/F$

For beam or shell element under bending load, U is to be derived by integration as the displacement varies along length. Take a length dx at distance x from one end, let M be the moment at dx due to load F.

$$U = \int_0^L \frac{M^2}{2EI} dx$$

For single point load W, calculate moment M in two parts either side of load at dx at a distance x from one end and calculate U in each part using above equation, integrate with limits 0 to $L/2$ and $L/2$ to L, and add both to get U in the beam for load W. Deflection (d) can be calculated under the load from equation $2U/W$. For multiple loads and UDL, U can be calculated but deflection cannot be calculated. But using Castgliano's theorem deflection can be calculated for any element, at any point and any number and type of loads.

3.4.4 Castgliano's Theorem

Statement: partial differentiation of the strain energy calculated as explained in 3.4.3 with respect to any one of static load (W) is equal to the displacement at the point and direction of W, similarly rotation for moment load.

For displacement at point where there is no applied load, assume load/moment, derive equation for displacement/rotation, equate assumed load or moment to zero to get required displacement or rotation. This theory can be applied to curved beams and applied to straight and bend pipe elements in *chapter 9*.

3.5 LONG SHELLS WITH THREE SUPPORTS

Long shells with three supports as shown in Figure 3.7 are a case of continuous beam which can be solved by theorem of three moments (Clapeyron's theorem). *A*, *B*, and *C* are supports (B middle). Moments at *A and C* are calculated from the loading diagram. Calculate the slope at *B* from both sides by the integration method and equate both (different signs) to get an equation from which unknown *M* at *B* can be calculated. Then by static equilibrium, three support reactions can be calculated.

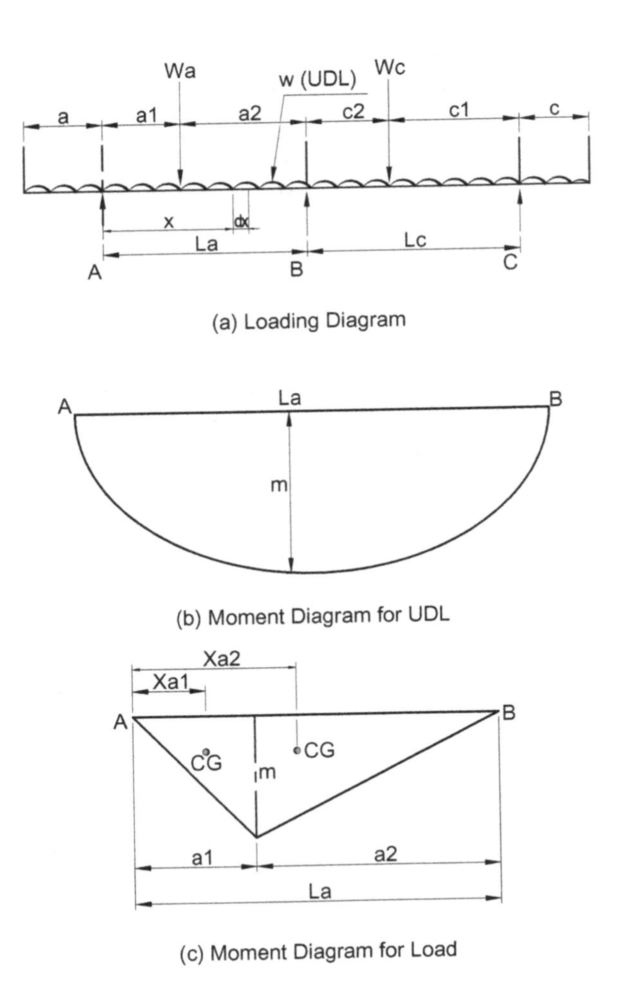

FIGURE 3.7 Long shell with three saddles.

M_a, M_b, and M_C = moments at support A, B, and C
L_a = length *AB* and L_c = length CB
I_a and I_C = second moment of area of the cross-section of shell *AB and CB*

To calculate slope at *B*, use the deflection equation given below at a distance x from *A* in span *AB*.

$$EI\, d^2y/dx^2 = M_a + x/L_a(M_b - M_a) + M$$

M is moment at a distance *x* from *A* in span *AB* due to the loading diagram of the span with ends *SS* and called free bending moment. Multiplying by *x* and integrating, we get

$$EI\, x\, dy/dx = EI\left[x\, i_x\right] = \int_0^{L_a}\left[M_a\, x + \frac{x^2}{L_a}(M_b - M_a) + M\, x\right]dx$$

Where i = slope, integrating and applying limits

$$EI[L_a\, i_b - 0\, i_a] = [M_a\, L_a^2/2 + \frac{L_a^3}{3L_a}(M_b - M_a) + \int_0^{L_a} M\, x\, dx$$

$$EI\, i_b = \left[\frac{L_a^2}{6}(M_a + 2M_b) + \Sigma A\, x'\right]/L_a$$

Where $\int M\, dx = \Sigma A\, x'$ = moment of the area (A) of the free bending moment diagram of *AB* at support *A*. x' = distance of CG of the area from support *A*.

Similarly, we get slope at *B* from the *C* side and by equating these two slopes, assuming $I_a = I_C$, we get an equation

$$\left[\frac{L_a^2}{6}(M_a + 2M_b) + \Sigma A\, x'\right]/L_a = \left[\frac{L_c^2}{6}(M_c + 2M_b) + \Sigma A\, x'\right]/L_c$$

Multiplying by six and simplifying, we get

$$M_a/L_a + 2M_b/(L_a + L_c) + M_c/L_c - 6(\Sigma A_a\, x_a/L_a + \Sigma A_c\, x_C/L_c) = 0 \qquad (3.18)$$

From Eq. 3.18, *Mb* can be calculated.

$$\text{If } I_a \neq I_C, \frac{Ma}{L_a I_a} + \frac{2M_b}{L_a\, I_a + L_c\, I_C} + \frac{M_c}{L_c\, I_C} - 6\left(\Sigma\frac{Aa\, xa}{L_a\, I_a} + \Sigma\frac{A_C\, x_C}{L_c\, I_C}\right) = 0$$

If the supports sink nonuniform, an additional factor is added as below

If support *A*, *B*, and *C* sink by $y_a\, y_b\, y_C$, add $6EI[(y_b - y_a)/L_a + (y_b - y_C)/L_C]$ in Eq. 3.18 in LHS

Note that the vessel with more than two supports is difficult to install that all the supports are positioned at the correct level to have reactions as per the calculation. Actual reactions will never be as per calculations, and it is advised to increase about 10–20% for further calculations. Further if the shell is made up with more than one part, their joint shall be rigid not to have relative rotation for validity of the above equations.

Example 3.4: Calculate reactions and moments in the shell of 16 m long with three supports as shown in Figure 3.7

UDL of shell weight w = 1000 kg/m, conc. Loads W_a = 1000 kg, W_b = 2000 kg
Dimensions in m: a = 2, a_1 = 2, a_2 = 4, c_2 = 2, c_1 = 3, c = 3

$$L_a = a_1 + a_2 = 6, L_C = c_1 + c_2 = 5$$

By beam formulas, M_a and M_c are:

$$M_a = w a^2/2 = 2000 \text{ and } M_c = w c^2/2 = 4500$$

For UDL: moment $m = w\ L_a^2/8$
Area of moment diagram (Figure 3.7b) $A_a = 2/3(m\ L_a) = w\ L_a^3/12$

$$A_a\ x_a/L_a = (wL_a^3/12)(L_a/2)/L_a = wL_a^3/24 = 9000$$

Similarly, $A_c\ x_c/L_C = w\ L_C^3/24 = 5208$
For concentrated load, W_a moment is given by

$$m = W_a\, a_1\, a_2/(4L_a)$$

Areas of moment diagram (Figure 3.7c) $A_{a1} = m\ a_1/2$, $A_{a2} = m\ a_2/2$

$$x_{a1} = 2a_1/3, x_{a2} = a_1 + a_2/3$$
$$[A_a\ x_a/L_a] = (A_{a1}\ x_{a2} + A_{a2}\ x_{a2})/L_a = m[(a_1/2)(2a_1/3) + (a_2/2)(a_1 + a_2/3)]/L_a$$
$$= W_{a\,a1}(2_{a1}^{\ 2} + 3_{a1.a2} + a_2^{\ 2})/(6L_a^{\ 2}) = 1778$$

Similarly, C side = 3200,
Rearranging Eq. 3.18 and substituting the above values

$$M_b = [-M_a L_a - M_C L_C + 6(A_a\ x_a/L_a + A_C\ x_C/L_C)]/[2(L_a + L_C)]$$
$$M_b = [-2000 * 6 - 4500 * 5 + 6(9000 + 1778 + 5208 + 3200)]/[2(6 + 5)]$$
$$= 3664 \text{kgm}$$

Reactions R_a and R_c can be calculated by taking moments at B from either side

$$R_a = [w(a + L_a)^2/2 + W_{a\,a2} - M_b]/L_a = 5389 \text{ kg}$$
$$R_C = [w(c + L_C)^2/2 + W_C c_2 - M_b/L_b] = 6467 \text{ kg}$$
$$R_b = \text{total load} - R_a - R_C = (1000{*}16 + 1000 + 2000) - 5387 - 6467 = 7143 \text{ kg}$$

3.6 LOADS

The loadings to be considered in designing other than pressure are

- D-Dead
- L-Live
- W-wind
- E-earthquake/seismic
- S-snow
- T-thermal, applied displacements

Loads due to mass (dead, live, and snow) are ultimately reacted by earth. Loads arising without mass are balanced by elasticity of material (wind, seismic, and thermal) or effects on earth at certain point of equipment in contact, an equal effect of opposite sign is induced at some other point of equipment in contact so that the resultant effect on earth and equipment is nil. Thus, only loads due to mass are transferred and reacted by earth. All mass less loads are not transferred to earth.

Load combination: Because all the loads will not act simultaneously, standards and codes consider the same with different procedures for analyzing stresses and coincident allowable stress (S).

$$\text{Sustained} + \text{occasional (wind or seismic or snow)} : 1.33\text{S (allowed)}$$

Method of loading: The mechanical properties of a material such as ultimate strength and elastic limit are usually determined by laboratory tests. To apply these results in engineering design, we require an understanding of the effects of many variables including the method of loading.

The method of loading affects the behavior of bodies under stress. There are many ways in which load may be applied to a body, but for most purposes, it is sufficient to distinguish the types of loading as under.

1. Short-time static loading as in material testing in universal testing machines.
2. Long-time static loading as in pressure vessels where pressure is raised gradually and is maintained for a longer time. Most materials will creep or flow to some extent and eventually fail under such pressure and temperature beyond its creep range.
3. Repeated loading: Load is applied and wholly or partially removed or reversed repeatedly. In this type of loading, the body fails to give desired service if high stresses are repeated for a few cycles even though stresses are within the allowable limit or if relatively low stresses are repeated many times.
4. Dynamic loading as in the effect of wind or seismic.

3.7 STRESSES

Basic stresses are direct (tensile or compression) and shear.

3.7.1 Tensile

Tensile stress is equal to force divided by the effective area of cross-section of plane where stress is desired perpendicular to the force direction. Practically, force is not uniform over the area at the point of application. It will become uniform after certain length. This transition length with nonuniform stress will not lead to plastic deformation of elemental areas with higher force due to the resistance offered by elemental areas with lower force. Plastic elongation will not start until all the elements of areas reach the yield point. At the point of boundary, practically local or discontinuity stresses exist which is normally bending and shear.

3.7.2 Compression

Pure compression exists only when the cross-section of plane perpendicular to the force, on which force is applied, is large enough compared to the length in the direction of force. An example is the stress induced in a washer when a bolt and nut is tightened with a washer in it. Deviation occurs when the length is increased and the body buckles and deflects in a direction perpendicular to the force in middle or at a plane with the lowest area. Such yield or plastic deformation occurs below the yield stress of the material.

3.7.3 Shear Stress

Shear stresses are pure, flexural (transverse), and torsional.

3.7.3.1 Pure Shear

Pure shear exists only when the distance between resistance and shear force plane tends to zero. As the distance increases, bending moment exists due to the increase in lever arm, and transverse shear and bending moment are induced. The example of pure shear stress is in rivets, bolts, etc. Torsional shear exists in rotating parts.

3.7.3.2 Transverse Shear Stress

Transverse shear stress in beams under bending is difficult to feel. Shear stresses on one side of an element are accompanied by shear stresses of equal magnitude acting on perpendicular faces of an element. Thus, there will be horizontal shear stresses between horizontal layers (fibers) of the beam, as well as transverse shear stresses on the vertical cross-section. At any point within the beam, these complementary shear stresses are equal in magnitude.

A single bar of depth $2h$ is much stiffer that two separate bars each of depth h. Sliding takes place at the contact area between two. This shows the existence of horizontal shear stresses in a beam. See Figure 3.8

To derive the equation for beam, assume that shear stress is the same at all points on width. Consider small area (dA) length dx and height dy at a distance y from NA as shown in Figure 3.9. There is a difference between moments of both sides which will result in balance force ($\sigma\, b\, dy$). For horizontal (length) equilibrium, this balance

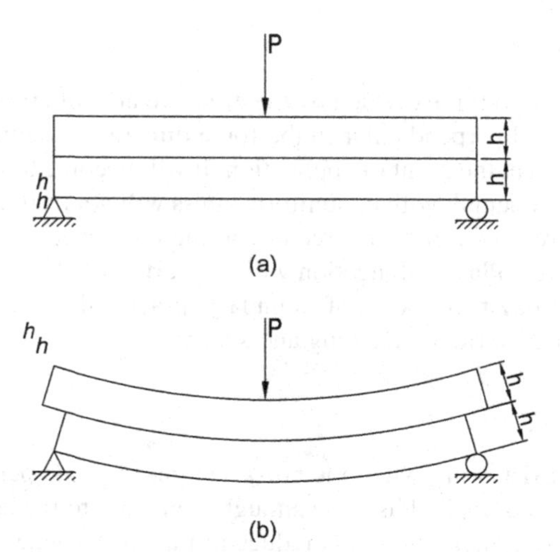

FIGURE 3.8 Horizontal shear stress in the beam.

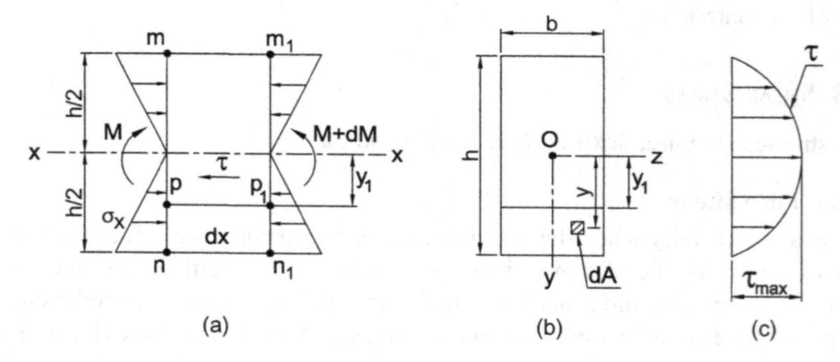

FIGURE 3.9 Shear stress in a rectangular beam.

force is resisted by force ($\tau\ b\ dx$) due to shear stress and equating both shear stress is given by Eq. 3.19.

$$\tau = \frac{\sigma b\, dy}{b\, dx} \tag{3.19}$$

where

σ = difference between bending stress at either side of dx

b = width of the beam at the point

Using equations $\sigma = dM\ y/I$ and $V = dM/dx$, Eq. 3.19 can be written as

$$\tau = \frac{(dM\ y/I) b\, dy}{b\, dx} = \frac{dM}{dx}\frac{y\, b\, dy}{I\, b} = \frac{V}{I\, b}\int (y\, b\, dy) = \frac{V Q}{I\, b} \tag{3.20}$$

where
V = shear force at section containing the point
Q = first moment of area = $\int(y\ b\ dy)$

The above integration is from point to top (or bottom) of the moment (lever arm y) at (NA-Neutral Axis of the total section) of the area (b dy).

Q for any point of the regular section = area from the line parallel to NA to top (or bottom) of section multiplied by the distance of line from NA.

Stress is zero at top and bottom and maximum at NA.

Example 3.5: Calculate shear stress in given rectangle sides b × h = 100 × 200 mm and V = 100 KN

For rectangle $I = b\,h^3/12$, Q at middle (NA) = $(b\,h/2)(h/4) = b\,h^2/8$

Substituting I and Q in Eq. 3.20,

$$\tau = \frac{V(b\,h^2/8)}{b^2h^3/12} = \frac{1.5V}{b\,h}$$

$$\tau = 1.5 * 1000/(100 * 200) = 7.5\,\text{MPa}$$

The general equation for max transverse shear stress is

$$\tau = \alpha\,V/A \tag{3.21}$$

where
V/A = average shear stress.
α = shape factor 3/2 for rectangle as derived in the above example.
For other sections, α can be derived as per above basics and given below.
4/3 for a solid bar and 2 for a thin cylinder

For ductile materials, shear stress is normally not significant compared to other stresses.

3.7.3.3 Torsional Shear

Torsional shear is given by a simple equation the same as bending stress, replacing torsion moment for bending moment and polar moment of inertia (J) for I.

$$J = I_{XX} + I_{YY}, \text{ for cylinder } J = 2I$$

3.7.4 Combined and Allowable Stress

Allowable stress is the max stress as calculated for the expected conditions of service and is less than the damaging stress because of uncertainty as for the conditions of service, nonuniformity of material, and inaccuracy of assumptions made in the stress analysis.

Allowable stresses as per IS800 are yield stress (Y) / factor of safety and given below for all types of stresses.

- Tensile axial $\sigma_t = 0.6Y$
- Tensile bending $\sigma_{bt} = 0.66Y$,

- Shear $\tau = 0.45Y$,
- Compression axial $\sigma_C = 0.6f.Y$,

where

$f = f_{CC}/(f_{CC}^n + Y^n)^{1/n}$, $f_{cc} = \pi^2 E/k^2$, $k = L/r$,

L = equivalent length = $K L_a$, L_a = actual length

The K value depends on boundary conditions [translation (T) and rotation (R)], zero (restrained) or free, is given in Table 3.3.

Allowable stresses for combined stresses are

$$\sqrt{\left[\sigma_t^2 + 4\tau^2\right]} < 0.6Y$$

$$\sqrt{\left[\sigma_b^2 + 4\tau^2\right]} < 0.9Y$$

$$\left[\sigma_C/(0.6f\ Y) \text{ or } \sigma t/0.6Y\right] + \sigma_b/0.66Y < 1$$

3.7.5 Compression Bending

Compression bending is the same as tensile bending provided the section is laterally supported for torsion buckling for sections such as I-beams, T-sections, angles, and channels. Else allowed stress is less and depends on *D/T* and *L/r* ratios. IS-800 provides the tables for allowed stress.

3.7.6 Fatigue

Practically, all materials will break under numerous repetitions of a stress that is not as great as the stress required to produce immediate rupture. This phenomenon is known as fatigue. Fatigue life or the number of cycles is inversely proportional to the stress ratio. The lower the induced stress, the higher the fatigue life. Details are covered in Chapter 9.

TABLE 3.3
K value for Boundary Conditions of a Beam

	A end		B end		
S.No	**T**	**R**	**T**	**R**	**K value**
1	0	0	0	0	0.65
2	0	0	0	free	0.8
3	0	free	0	free	1
4	0	0	free	0	1.2
5	free	0	free	0	2
6	0	0	free	free	2

T = Translation, R = Rotation

3.7.7 Brittle Fracture

Brittle fracture is a term applied to an unexpected brittle failure of a ductile material at low temperatures, where large plastic strains are usually noted before actual separation of the part. Major studies of brittle fracture started when failures such as those of welded ships operating in cold seas led to a search for the effect of temperature on the mode of failure. For a brittle fracture to take place, the material must be subjected to a tensile stress at a location where a crack or other very sharp notches or defects are present, and the temperature must be lower than the so-called transition temperature.

3.7.8 Stress Concentration

The distribution of elastic stress across the section of a member may be nominally uniform or may vary in some regular manner, as illustrated by the linear distribution of stress in flexure. When the variation is abrupt so that within a very short distance, the intensity of stress increases greatly, the condition is described as stress concentration. It is usually due to local irregularities of form such as small holes, screw threads, scratches, and similar stress raisers. The ratio of this increased actual stress divided by stress as per normal elastic theory is the stress concentration factor. It can be better understood by an example. Consider a straight rectangular beam, originally of uniform breadth b and depth D, which has cut across the lower face a fairly sharp transverse V-notch of uniform depth h, making the net depth of the beam section at that point $D–h$. If now the beam is subjected to a uniform bending moment M, the nominal fibre stress at the root of the notch may be calculated by ordinary flexure formula $\sigma = M_c/I = 6M/[b(D–h)^2]$. But the actual stress σ' is very much greater than this because of the stress concentration that occurs at the root of the notch. The ratio σ'/σ actual stress divided by nominal stress is the stress concentration factor.

3.8 ANALYSIS METHODS

Stress analysis of pressure vessels can be performed by an analytical method using mathematical solutions directly or by integration based on applicable theory. Mathematical solution includes available formulas, and combination of them, exploring them to make new formulations using fundamentals by making acceptable assumptions. Available formulas include empirical formulas based on practical and experimental data such as the frictional factor, combination of dimensionless numbers to calculate the heat transfer coefficient, etc. When the problem is too complex by this method, finite element analysis is adopted (refer Part 5 of ASME S VIII D 2). If the problem is beyond these solutions or alternate to finite element analysis, a proof test can be carried out [refer UG-19(c) and UG-101 of ASME S VIII D 1]. Some analysis methods are given below. Others are used where ever is required in this book (refer sections 9.6.2.2 and 9.7) for the analysis method using Castigliano's first theorem).

3.8.1 Equilibrium Method

The equilibrium method is by forming equations using static equilibrium of any free body diagram of element like equating forces in each direction and taking moment at any point in each axis. $\Sigma F = 0$, $\Sigma M = 0$.

3.8.2 Integration Method

The integration method is the summation method of all elements of the body to calculate any mathematical equation by taking the infinitesimal element, when a direct formula is not available. This method is used for calculating the equation for MI of a section of body at its NA in 3.1.9 which is very simple. Similarly for any complicated problem, this method will give solution. It can involve double or triple integration for two- or three-dimensional problems.

3.8.3 Theorem of Least Work

When an elastic system is statically loaded, the stresses are distributed in such a way to make the strain energy minimum consistence with equilibrium and the boundary conditions. It is used to solve statically indeterminate problems. Select unknown like boundary reactions to be considered as redundant until it is determinate. The strain energy of the system is then formed, and its partial derivative with respect to each of the redundant reactions or stresses is then set to zero, and the resulting equations are solved for redundant reaction or stress.

3.8.4 Experimental Method

The experimental method is used by carrying performance analysis of actual or prototype element. The values of input and output are tabulated or graphed so as to use for all input values of identical or similar element. On the basis of the analysis, empirical formulas can be derived.

3.8.5 Steps in Mechanical Design of Equipment

Divide the equipment into structural members. Analyze the end members first with only loads on it other than pressure by making the free body (loading) diagram. Calculate the maximum stress including stress due to pressure and compare with allowed limit. Transfer the loads to a common node point with adjacent members. Continue until all elements are analyzed and reactions and moments on concrete pedestal or supporting structure are calculated. Practically, the path is not a line. Boundary conditions of edges of members are assumed. Certain members are so stiff that they only transfer the forces and need not be checked for stresses.

A structural member may be of such a form or may be loaded in such a way that the direct use of formulas for the calculation of stresses and strain produced in it is ineffective. Then one must resort either to numerical techniques such as the finite element method or to experimental methods. Experimental methods can be applied to the actual member in some cases, or to a model thereof.

3.9 ELASTIC FAILURE THEORIES

σ_1, σ_2, and σ_3 are principal stresses; Y yield stress.

3.9.1 Maximum Principal Stress Theory

This theory states that elastic failure (yield) occurs when the maximum principal stress becomes equal to yield stress. It disregards the effect of other two principal stresses.

$$Y = \max(\sigma_1, \sigma_2, \sigma_3)$$

ASME S-I & S-VIII D-1 and IBR use this theory. It is oldest and most widely used. This theory is used for biaxial states of stress assumed in thin walled pressure vessels. It accurately predicts failure in brittle materials; it is not always accurate in ductile materials. Ductile materials often fail along lines 45° to the applied force by shearing, long before tensile or compressive stresses are maximum.

3.9.2 Maximum Strain Theory

This theory states that elastic failure occurs when the maximum tensile strain becomes equal to Sy/E.

3.9.3 Maximum Shear Stress Theory

This theory states that elastic yielding occurs when the maximum difference of shear stress becomes equal to $Y/2$. Max shear stress in a tensile test specimen is half the axial principal stress. It assumes that yielding starts in planes of maximum shear stress for biaxial state of stress when

$$(\sigma_1 - \sigma_2)/2 = Y/2$$

In tri-axial state of stress when

$$\max[(\sigma_1 - \sigma_2)/2, (\sigma_2 - \sigma_3)/2, (\sigma_3 - \sigma_1)/2] = Y/2.$$

This theory closely approximates experimental results. ASME S VIII D-2 & S III uses this theory.

3.9.4 Theory of Constant Energy of Distortion (Von-Mises Theory)

This theory states that elastic yielding occurs when distortion energy at a point in stressed and equal to distortion energy in a uniaxial test specimen at a point it begins to yield. The stress is called Von-Mises stress and given by

$$Y = \sqrt{\{(\sigma_1 - \sigma_2)^2 + (\sigma_2 - \sigma_3)^2 + (\sigma_3 - \sigma_1)^2\}/2}$$

ASME S-VIII D 2 Part 5 uses this theory.

3.10 MATRIX EQUATION OF THE BEAM ELEMENT

For the beam element, AB with one end A fixed, other end B is applied with force tensor 'F', no other loads in the element length, displacement tensor at B, [D] = [K] × [F], [K] = flexibility matrix [6 × 6] and force tensor [F] = [S] × [D], [S] = stiffness matrix.

For the cantilever beam, [K] and [S] can be derived from elastic theory and beam formulas and are given below.

$$[K] = \begin{matrix} L/AE & \nu L/AE & 0 \\ \nu L/AE & L^3/3EI & L^2/2EI \\ 0 & L^2/2EI & L/EI \end{matrix} \qquad [S] = \begin{matrix} AE/L & \nu AE/L & 0 \\ \nu AE/L & 12EI/L^3 & 6EI/L^2 \\ 0 & 6EI/L^2 & 4EI/L \end{matrix}$$

Refer Chapter 9 for a general 6 × 6 flexibility matrix which is used in the finite element method with beam elements.

REFERENCE

1. IS-800

4 Pressure Vessel Design Basics

4.1 GENERAL

Notation: P – Pressure, t – Thickness, σ – Stress

1. *Pressure vessels:* Pressure vessels are the containers of fluid static or flowing, internal and/or external under internal and/or external pressure.
2. *Closed pressure vessel*: Closed vessels containing fluid under pressure are termed closed pressure vessels. Boilers and their components and heat exchangers with water or steam as the internal fluid (water tubes, steam drum, etc.) or external (fire tube, furnace tube, etc.) or both sides (tubes in steam feed water heaters) are specific types of pressure vessels.
3. *Open vessel*: It is defined as the enclosure open to the atmosphere of spherical, cylindrical, rectangular, or other geometry or in combination, holding any material or fluid such as tanks, bunkers, chimneys, etc, which are under pressure due to static head or flow.

4.1.1 Shape of the Shell

The shell is the major part of a vessel enclosing all the internals required for the given process and allowing external attachments connecting other equipments including supports, instruments, controls, etc. Internal pressure is the major load for a shell.

Shapes of shells are listed below:

- Spherical
- Cylindrical
- Conical
- Rectangular
- Toroidal (Torus)

The factor for selecting the shape of a vessel is primarily the process requirement; however, comparison of main shells and various other factors in the selection of the shape of the shell is tabulated in Table 4.1.

4.1.2 Pressure and Static Head

To obtain a physical feeling of pressure, consider being submerged in water at a particular depth. The force of the water one feels at this depth is a pressure, which is due to static head.

DOI: 10.1201/9781003091806-4

TABLE 4.1
Comparison of Shells

	Spherical	Cylindrical/cone	Rectangular
Utilization of material	100%, maximum	Intermediate	Minimum
Internal space utilization	Minimum	Intermediate	Maximum
Weight of shell	Minimum	Intermediate	Maximum
Local stresses	Maximum	Intermediate	Minimum
Utilization of external space	Maximum	Intermediate	Minimum
Supporting arrangement	Complicated	Moderate	Simple
Pressure application	Any	Any	Low

Atmospheric (absolute) pressure is due to static head of air from the altitude where it is measured to the height where the atmosphere ends. At the mean sea level, the atmosphere exists up to height (H = 12 Km approximately).

Air density is 1.3 kg/m³ at the mean sea level, zero at height *H*, and average 0.866 kg/m³.

$$\text{Pressure due to static head is } 12000 \times 0.866 = 10400\,\text{kg/m}^2 = 1.04\,\text{kg/cm}^2$$

This pressure is the reaction of weight (1.04 kg) of the air column over an area of sq.cm.

4.2 LOADS

The loadings to be considered in designing pressure vessels in addition to loads given in 3.3 are:

- Internal and external design pressure (P) including static head (P_s).
- Weight of the vessel, internal fluid, and components under operating or test conditions (D).
- Live loads (L) and superimposed static reactions from weight of attachments such as motors, machinery, piping, linings and insulation, and vessel supports.
- Cyclic and dynamic reactions due to pressure or thermal variations or from equipment mounted on a vessel and mechanical loadings.
- Wind (W), snow(S), and seismic (E) reactions, where applicable.
- Impact reactions such as those due to fluid shock.
- Thermal loadings due to temperature gradients and differential thermal expansion (T).
- Abnormal pressures, such as those caused by deflagration.
- Test pressure and coincident static head acting during the test.

Pressure (mass less) load is reacted by internal energy of the element material, and no support is required (no load on earth).

Load combination: Hoop stress in the cylindrical shell under internal pressure is main load. Other loads induce stresses which generally will not combine with hoop stress, but combine with longitudinal stress which is half of the hoop stress. Further allowable stress for the combined loads is up to three times the tensile stress.

Unless stresses due to combination of loads are very high, hoop stress alone will decide the thickness of the element.

Load combination is related to allowable stress. In pressure part load combination, the primary load is always pressure. Whenever one or more other loads are acting, the allowable stress will be greater than allowed tensile stress, and the value depends on the stress category of other stresses induced due to other loads. Some normal combination loads and their stress category as per ASME S VIII D 2 are

$$P + P_s + D + L = 0.9P + P_s + D + L + S + (0.6W \text{ or } 0.7E) = P_m + P_b = 1.5S$$

$$P + P_s + D + L + T = 0.9P + P_s + D + 0.75(L + T) + 0.75S = P_m + P_b + Q = 3S$$

4.3 MEMBRANE THEORY

4.3.1 Shells of Revolution

A line straight or profile (A-B) is revolved around an axis (YY) to form a shell which is called shell of revolution and axis *YY* is called axis of revolution as shown in Figure 4.1. The following are examples.

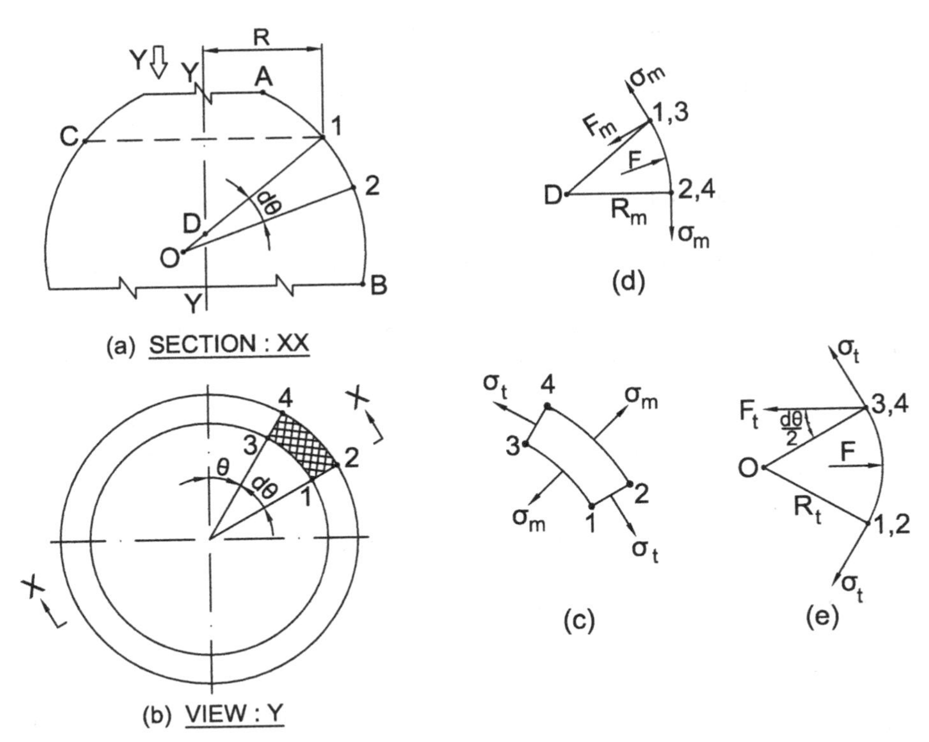

FIGURE 4.1 Partially spherical membrane shell.

1. *Cylindrical shell*: line (middle of thickness) straight and parallel to axis
2. *Conical*: inclined line at an angle (α half cone angle) to the axis.
3. *Spherical*: semicircular line touching the axis of rotation.
4. *Partially spherical* (ellipsoidal or torispherical): line of relevant profile touching the axis of rotation.
5. *Toroidal*: line is a closed curve, usually a circle at a radius from axis of rotation.

If the thickness of the line of shell of revolution is less than 1/10th of minimum dimension of shell and properly shaped and correctly supported to retain the shape of the shell under uniform pressure load, the shell is called membrane shell as only membrane stresses are induced. Shells with a larger thickness are called thick shells and covered in Chapter 5.

4.3.2 Membrane Equation

Figure 4.1 shows the mean profile line in (a) of a partially spherical membrane shell under uniform internal pressure.

Notation (see Figure 4.1):

F = Force on element 1234

R = Mean radius normal to axis of rotation at point 1 on the shell surface

t = thickness of head

R_m = actual (meridian) radius of the curve at any point (0–1)

L = Length

R_t = tangential radius or normal distance from the point to the axis of rotation (1-D).

m, t = suffix for meridian (long), tangential (circ)

For membrane shells other than cylindrical shells, meridian and tangential are used in place of longitudinal and circumferential due to curvature in both directions. A line (profile AB) that represents the intersection of the wall and a plane containing the axis of the vessel is called a *meridian*, and a line (1C) representing the intersection of the wall and a plane normal to the axis of the shell is called *tangential*. Obviously, the meridian through any point is perpendicular to the circumference through that point.

Consider a differential shell element 1234 at latitudinal radius R subtend angle $d\theta$ at an angle θ as shown in figure. If the loading is axisymmetric like uniform internal pressure, boundary forces of element are membrane stresses only σ_m and σ_t as shown in Figure 4.1.

By equilibrium in the direction normal to the shell element (direction F in figure), Eq. 4.1 is derived.

Pressure force F normal to element 1234

$$F = P\,L_{12}\,L_{13}$$

The component of resisting force due to σ_m on element side 13 and 24 in the direction of F

$$F_m = \sigma_m L_{13} t \sin\left(\frac{d\theta}{2}\right) + \sigma_m L_{24} t \sin\left(\frac{d\theta}{2}\right)$$
$$= \sigma_m (L_{13} + L_{24}) t \sin\left(\frac{d\theta}{2}\right)$$

Similarly, the component of resisting force due to σ_t on element side 12 and 34 in the direction of F_t

$$F_t = \sigma_t L_{12} t \sin\left(\frac{d\theta}{2}\right) + \sigma_t L_{34} t \sin\left(\frac{d\theta}{2}\right)$$
$$= \sigma_t (L_{12} + L_{34}) t \sin\left(\frac{d\theta}{2}\right)$$

Equating forces

$$F = P L_m L_t = F_m + F_t = [\sigma_m (L_{13} + L_{24}) + \sigma_t (L_{12} + L_{34})] t \sin\left(\frac{d\theta}{2}\right)$$

Denoting L_m for L_{13} and L_{24}, L_t for L_{12} and L_{34} and $\sin\left(\frac{d\theta}{2}\right) = \frac{L_t/2}{R_m} = \frac{L_m/2}{Rt}$

$$P L_m L_t = 2\sigma_m L_m t \frac{L_t}{2R_m} + 2\sigma_t L_t t \frac{L_m}{2R_t}$$

Simplifying, *membrane equation* 4.1 is obtained.

$$\frac{P}{t} = \frac{\sigma_m}{R_m} + \frac{\sigma_t}{R_t} \tag{4.1}$$

By equilibrium of pressure force and component of force due to σ_m in the direction of the axis of rotation using the resisting thickness component = $t \sin\theta$

$$\pi R^2 P = (2\pi R)(t \sin\theta)\sigma_m$$
$$\sigma_m = P/2t(R/\sin\theta),$$

Substituting $R/\sin\theta$ for R_t

$$\sigma_m = \sigma_L = \frac{PR_t}{2t} \tag{4.2}$$

Substituting σ_m from Eq. 4.2 in Eq. 4.1, σ_t can be calculated and depends on shell geometry. σ_m, σ_t, and σ_r are called principal stresses. Principal stress is defined as stress in a plane in which shear stress is zero.

The third principal stress is radial stress maximum at inside P and 0 inside; *taking* average value $\sigma_r = P/2$

Membrane theory neglects radial stress as its value is very less compared to σ_m and σ_t. Membrane stresses for different shells are computed from Eq. 4.1 and 4.2.

Cylindrical shell: Substituting $R_m = R_L = \infty$, and $R_t = R$ in Eqs. 4.1 and 4.2

$$\sigma_t = \sigma_C = \frac{PR}{t} \ \& \ \sigma_m = \sigma_L = \frac{PR}{2t}$$

Conical shell: Substituting $R_m = R_L = \infty$, and $R_t = R/\cos\alpha$ in Eqs. 4.1 and 4.2

$$\sigma_t = \frac{PR}{t\cos\alpha} \ \& \ \sigma_m = \frac{PR}{2t\cos\alpha}$$

Spherical shell: Substituting $R_m = R_t = R$, in Eq. 4.1

$$\sigma_t = \sigma_m = \sigma_C = \sigma_L = \frac{PR}{2t}$$

Membrane stresses in other shells under uniform internal pressure are covered in Chapter 5.

4.4 STRESSES

Stresses due to internal pressure are a major part in pressure vessels. The tensile stress in thin (R/t > 10) membrane shells due to pressure is considered the same across thickness and called membrane stress.

Due to internal pressure, the element of vessels is subjected to three principal stresses

1. *Circumferential/hoop*: approximately equal to $p\ R/t$, maximum inside and minimum outside.
2. *Longitudinal*: approximately equal to $p\ R/2t$, maximum inside and minimum outside.
3. *Radial*: equal to pressure inside and zero outside.

The difference of stress from inside to outside decreases with the increase in the R/t ratio and is insignificant above 10. The radial stress above this ratio is very less compared to other two stresses and can be neglected. Codes give a slightly modified formula for stresses (average) to cover most of the practical shells up to $R/t = 2$. Almost all applications, the R/t ratio of shells will be above this value.

The circumferential stresses are applicable for the elements remote from discontinuity due to branch, end closing part, fitting, etc. The stress changes local to discontinuity from a distance of $\sqrt{(R\ t)}$. The local elements of shell and joining part at the discontinuity share the stress proportional to their stiffness. The stress in the vessel joining any type of end closing part reduces, and that of the end closing part increases.

4.4.1 Stress Categorization

4.4.1.1 Primary Stresses

A tensile, compressive, or shear stress is developed by imposed loading which is necessary to satisfy the law of equilibrium of forces and moments. These are not self-limiting.

1. *Primary general*: Primary general stresses are generally due to internal or external pressure or produced by sustained external forces and moments.
   σ (direct) = F/A
   τ (shear) = F/A
   τ_t (torsional shear) = M/J
   σb (bending) = M/Z
2. *Primary general membrane stress, P_m*: this stress occurs across the entire cross section of the pressure vessel. It is remote from discontinuities such as head–shell and cone–shell intersections, nozzles, and supports.
3. *Primary general bending stress, P_b*: these stresses are due to sustained loads and are capable of causing collapse of vessels. Examples are bending stress in center of a flat head or crown of a dished head and bending stress in the ligaments of closely spaced openings.
4. *Primary local membrane stress, P_L*: Local membrane stresses are produced from sustained loads (by pressure alone or by other mechanical loads) and are limited to a distance $\sqrt{(R\ t)}$ in meridian direction. When these stresses reach yield, the load is redistributed to a stiffer portion of the vessel and is unrelenting. Examples are membrane stresses from local discontinuities such as junctures of head-shell, cone-cylinder, nozzle-shell, nozzle-flange, head-skirt, and shell-stiffener ring and from local sustained loads such as support lugs, nozzle loads, beam supports, and attachments.

4.4.1.2 Secondary Stresses

These are self-limiting, strain induced. Local yielding and distortions will reduce the stress. Examples are thermal and bending stresses at a gross structural discontinuity.

1. *Secondary membrane stress, Q_m*: examples are thermal, axial stress at the juncture of a flange and hub of the flange, and membrane stress in the knuckle area of the head.
2. *Secondary bending stress, Q_b*: examples are discontinuity stresses at stiffening or support rings, and shell–nozzle juncture.

4.4.1.3 Peak Stress F

Examples are stresses at corner of a discontinuity, thermal stresses in a wall caused by a sudden change in the surface temperature, thermal stresses in a cladding or wall overlay, and stress due to the notch effect (stress concentration).

4.4.2 Shear Stress in a Cylindrical Shell

Shear stress in shells under bending is max at middle and is given by Eq. 3.20 = V Q/(I b) where

V = shear force, d_O and d_i are outside and inside diameters

Q = area from top to centerline × distance middle to its CG

$$= \frac{\pi d_O^2}{8}\frac{2d_O}{3\pi} - \frac{\pi d_i^2}{8}\frac{2d_i}{3\pi} = \frac{d_O^3 - d_i^3}{12}$$

Moment of inertia $I = \pi(d_O^4 - d_i^4)/64$, b = 2 × thickness = ($d_O$–$d_i$)

Substituting

$$\text{Shear stress} = \frac{V}{b}\frac{Q}{I} = \left(\frac{V}{d_o - d_i}\right)\frac{(d_o^3 - d_i^3)/12}{\pi(d_o^4 - d_i^4)/64}$$

For thin cylinders with mean radius (r)

$$r\,t = (d_o^2 - d_i^2)/4,\ r^2 t = (d_o^3 - d_i^3)/24,\ r^3 t = (d_o^4 - d_i^4)/64$$

Using the above equations, shear stress

$$\tau = V/(\pi r\,t) \tag{4.3}$$

4.4.3 Allowable Stresses

Allowable stresses:

Allowable stress (*S*) refers to basic allowable stress for tensile or general membrane stress.

S = min of Sy/(1.5 to 1.6) or St/(2.7 to 3.5) below the creep temperature
S = min of Sy/(1.5 to 1.6) or St/(2.7 to 3.5) or Sr/(1.25 to 1.50) or Sc, above the creep temperature

where

S_Y = Yield strength at metal temperature (*T*)
S_t = Ultimate tensile strength (UTS) at ambient temperature
S_r = Average stress to produce rupture in 100,000 hours at T
Sc = Average stress to produce an elongation of 1% (creep) in 100,000 hours at temperature

The above varies marginally in different codes mainly on the factor of safety.

If codes do not specify allowable stress, values of S_y, S_r, and S_c at temperature are to be obtained specifically from material manufacturers. If the values are not available, specified UTS is used for computing the value of *S*. The Indian code for boilers IBR specifies S_y at elevated temperatures = $n\ S_t$ below the creep range.

Here the conservative value of *n* is given by

$$n = 0.4, 0.38, 0.36, 0.34, \text{ and } 0.32 \text{ for}$$
$$T = 250, 300, 350, 400, \text{ and } 450°C$$

Table 4.2 gives *St* at room temperature, and S at working temperature for commonly used materials. Other allowable stresses will vary and depend on the type of stress and loading and specifications along with the equations of respective design rules and covered in respective chapters. Some of the general rules are given below.

1. 1.2 times when combined with seismic/wind/snow loads or loads can be reduced by 1/1.2
2. 0.8 times for shear stress

Allowable stress for each type of stress or combination depends on whether primary or secondary, general or local, thermal, etc. and is normally as given below:

$P_m < S$ as the stress is uniform over thickness and due to sustained load
$(P_b \text{ or } P_L \text{ or } Q_m \text{ or } P_L + P_b) < SPL$ as the stress is local and/or primary bending

TABLE 4.2
Allowable Stresses (appx.) in Mpa at Temperature °C of Some Commonly Used C&LAS and SS Material

	UTS	**250C**	**275**	**300**	**325**	**350**	**375**	**400**	**425**	**450**
CS plates	420	118	117	115	112	108	104	89	76	63
CS plates	485	138	137	135	132	128	123	101	85	67
CS tubes	325	93	92.5	92	91	88	84	76	66	55
CS tubes & pipes	415	118	118	118	118	117	106	89	77	63
CS forgings	485	136	133	129	125	121	117	101	84	67
	UTS	**450C**	**475**	**500**	**525**	**550**	**575**	**600**	**625**	**650**
Gr11 plates	515	143	107	73	52	36	25	18	12	8
gr11-tubes & pipes	415	97	95	77	51	37	25	18	12	8
Gr22 plates	515	130	116	89	64	45	30	20	13	8
gr22-tubes & pipes	415	114	100	81	64	48	35	24	16	9.5
gr91-tubes&pipes	585	141	134	126	117	107	89	65	46	29
	UTS	**300C**	**350**	**400**	**450**	**475**	**500**	**525**	**550**	**575**
SS.18-8/304	515	116	111	107	103	101	99	97	93	80
SS.16-12-2Mo/316	515	119	114	111	108	108	107	106	105	100
SS18-10-ti/321	515	127	123	119	115	114	113	112	95	59
	600C	**625**	**650**	**675**	**700**	**725**	**750**	**775**	**800**	**825**
SS.18-8/304	65	51	42	33	27	21	17	14	11	8
SS.16-12-2Mo/316	80	65	50	39	30	23	18	14	11	8
SS18-10-ti/321	45	33	24	18.3	12.6	8.4	6.2	4.4	2.8	1.6

$(P_L + P_b + Q) < SPS$ as the stress is self limiting

$(P_L + P_b + Q + F) < (>3S)$

Peak stresses as specified in codes generally limited to 4S.

Where SPL = 1.5S and SPS = max $(3S, 2S_y)$

4.4.4 Thermal Stresses

Thermal stresses are developed whenever the expansion or contraction that would occur normally as a result of heating or cooling of component is prevented. These are "secondary stresses" because they are self-limiting. That is, yielding or deformation of the part relaxes the stress. Thermal stress will not cause failure by rupture in ductile material except by fatigue over repeated application. They can, however, cause failure due to excessive deformations. All most all equipments (except piping) are rigid between supports. Therefore, one support is designed as fixed and other sliding. If both supports are fixed, expansion joint (bellow) is provided between them. In such case, pressure thrust is induced in both fixed supports and generally not viable to design support. For low pressures, ties are provided between two parts separated by expansion joint to prevent supports from getting pressure thrust.

Piping systems are generally not rigid and providing one fixed and rest sliding will land in vibration problems. Several supports or restraints are required to prevent vibration. Pressure thrust is not high, and design of support or restraint is viable. A system can be designed to provide sufficient flexibility. Bellows are provided for low pressures and only when the required flexibility cannot be provided. Detailed analysis is covered in Chapter 9.

Effects of discontinuity, openings, and attachments to shell and external loads are covered in Chapters 7 and 8.

5 Internal Pressure

Notation: UTS – ultimate tensile stress, S – allowed stress, P – internal pressure, σ – stress, t – shell thickness

Suffixes: m – meridian, L – longitudinal, t – tangential, o – outside, and i – inside

5.1 CYLINDRICAL PARTS

Notation:

R = radius at the point across thickness

R_i and R_o = inside and outside radius

$k = R_o/R_i$

Suffixes: c = circumferential (circ) r = radial

The three principal stresses in cylindrical parts under internal pressure by Lame's theory are:

1. Circ or hoop stress = $\sigma_c = k_1 P\, R_i/t$
2. Long stress = $\sigma_L = k_2 P\, R_i/(2t)$, and
3. Radial stress = $\sigma_r = k_3 P R_i/t$

where

$$k_1 = \frac{(Ro/\mathrm{R})2+1}{Ro/\mathrm{R}+1}$$

$$k_2 = \frac{2Ri}{Ro+Ri}$$

$$k_3 = \frac{(Ro/\mathrm{R})^2-1}{Ro/\mathrm{R}+1}$$

5.1.1 Circumferential Stress

Circ stress is the maximum of the three principal stresses and varies across thickness as per the value of k_1, which is ≥1.

$$k_1 \text{ is maximum at inside and } = \frac{k^2+1}{k+1}$$

k_1 increases with the increase in thickness and is minimum at outside and equal to 1. For thin cylinders ($R_i/t > 10$), k_1 at inside is maximum 1.05 when $R_i/t = 10$. Therefore, stress can be considered constant across thickness with maximum 5% error. Maximum stress in the cylinder due to internal pressure is circumferential stress $k_1 P\, R_i/t$; therefore, min thickness required for given pressure and to limit stress to allowed (S) is given by Eq. 5.1

$$\mathrm{t} = \frac{\mathrm{P\,R_i}}{\mathrm{S/K_1}} \tag{5.1}$$

DOI: 10.1201/9781003091806-5

Because this thickness cannot be found without iteration, S/k_1 is replaced with a lesser value of S in various boiler and pressure vessel codes, with $S–0.6P$ in the code[1]. It is valid up to $t = R_i/2$ or $P = 0.385S$. At the limiting value, the code equation gives hoop stress almost the same as average stress across thickness by Lame's equation. The code equation gives a marginally higher thickness over Lame's equation up to the limiting condition and conservative.

Although S depends on the material and temperature, solving the code equation with S = 100 MPa and P = 38.5 MPa, t is equal to $R_i/2$. That is, both code limits match.

Applications more than the above pressure are very rare. Above the limiting condition, the code gives less thickness than Lame's equation, and the designer is advised to use the equation from Ref. 2, which is based on maximum shear stress theory. The Ref. 2 equation gives less thickness than that with the code.[1]

For low pressure (<2 MPa) or thin cylinder ($R_i/t > 50$), simple equation $t = (P/S)R_i$ gives less than 1% error over Lame's equation. However, the correct thickness can be found by replacing R_i by a higher radius little over the mean radius, mean for lowest pressure and higher and higher as pressure goes up.

5.1.2 Longitudinal Stress

Longitudinal stress is the same across thickness. The K_2 value is less than 1 and reduces as pressure or thickness increases.

Therefore, S/k_2 is replaced with a higher value of S, with $S + 0.2P$ in the code[1] and valid for thickness and pressure the same as in circumferential stress. Longitudinal stress is approximately half of σ_c. It is more than that with Lemi's formula up to the limiting pressure.

5.1.3 Radial Stress

Radial stress at inside is maximum and equal to P and zero at outside. It is much less than other two principal stresses and insignificant for thin cylinders.

5.2 SPHERICAL PARTS

Two of the principal stresses other than radial in the spherical shell or hemispherical dish under internal pressure are the same and approximately equal to σ_L of the cylinder with the same diameter. Thickness required is approximately half that of the cylinder for the same parameters and given by

$$t = \frac{0.5 P R_i}{S / K_1}$$

The code[1] gives the equation for calculating the thickness with limiting pressure equal to $0.665S$ as follows:

$$P R_i / \left[2\left(S - 0.1P\right)\right]$$

5.3 CONICAL PARTS

Notation (see also Figure 5.1)

R = mean radius perpendicular to axis
α = half cone angle
T_C = cone thickness

At any point in the cone, two principal stresses exist as per membrane theory: tangential and meridian. The actual radius at the point is equal to $R/\cos\alpha$.

The hoop stress is the same as that of the cylinder called equivalent cylinder with radius equal to $R/\cos\alpha$ and thickness required is calculated as an equivalent. However, it is applicable only remote from its junction from the shell. At junctions, *discontinuity stresses* are induced in addition to principal stresses in the cone, which are covered in Chapter 7. To understand the effect of discontinuity, consider the equilibrium of forces at the shell to cone junction as shown in Figure 5.1. The figure is drawn with the mean thickness and shows notation and forces.

For static equilibrium at point *A*

$$N = T\cos\alpha \text{ and } F = T\sin\alpha$$

Substituting $T = N/\cos\alpha$ and $N = P\,R/2$, the equation for F can be expressed as Eq. 5.2

$$F = P\,R\tan\alpha/2 \tag{5.2}$$

where F, T, and N are forces per unit circumference acting at A or B as shown in Figure 5.1.

At junction A, force F is compressive, and the cone tends to deflect inwards and the shell outwards under internal pressure. Reverse is the case at small end junction point B, and force F is tensile.

For $\alpha \le 30°$, a simplified analysis (reinforcement method) in lieu of *discontinuity stresses* can be used. This method is based on calculating the area required to withstand additional radial force F and by providing the area in both shells and cone locally at joint in addition to withstanding the hoop stress due to pressure and/or by adding a reinforcing pad at the junction.

Reinforcement method: The required area (A_r) is directly proportional to R and F (Eq. 5.2) and inversely to allowable stress S and given by

$$A_r = K\,F\,R/S$$

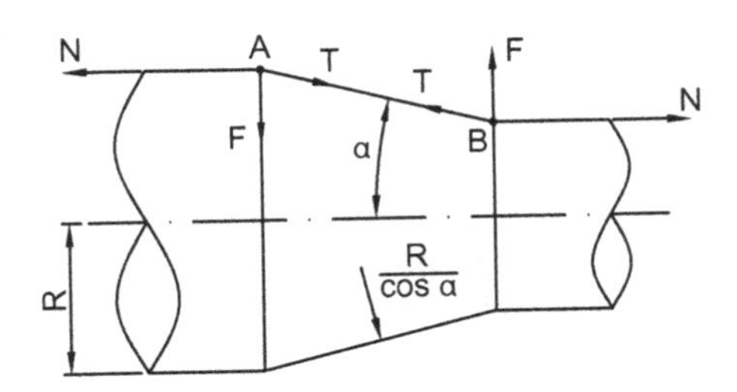

FIGURE 5.1 Conical reducer.

The code[1] gives a constant of proportion $K(<1) = 1-\Delta/\alpha$, where Δ = angle $\leq 30°$ which depends on the ratio of P/S for each joint at large and small ends of the cone and is given in Tables 1-5.1 and 1-5.2 of the code[1].

K = 1 if the cylinder length at the large end is less than $2\sqrt{(R\ t)}$ and less than $1.4\sqrt{(R\ t)}$ at the small end. Due to the insufficient length, the shell will not contribute reinforcement.

If $\Delta > \alpha$, K will become negative which indicates that the reinforcement area available in the cylinder and cone is adequate.

The available area is the area with net thickness in excess of max required thickness to withstand pressure, and limiting lengths are given below.

$$\text{Large end: shell} = \sqrt{RT},\ \text{cone} = \sqrt{RT_C/\cos\alpha}$$

$$\text{Small end: shell} = 0.78\sqrt{RT},\ \text{cone} = 0.78\sqrt{RT}/\cos\alpha)$$

The detailed calculation is illustrated by the example in Table 5.1.

TABLE 5.1

Compensation Calculation Due to Internal Pressure at Cone to Shell Junctions

***Input data*: pressure P = 1MPa, all. Stress S = 138MPa, E = 200000MPa**			**Large (L)**	**Small (s)**
Thickness of the cone (half cone angle α = 30)	T_C	mm	14	14
Thickness of cylindrical shells	T	mm	12	8
Mean radius	R	mm	1500	1000
Length of cylindrical shells	L_a	mm	300	300
Calculations:				
Required thickness of the cone for pressure = P R/(S cos α}	t_C	mm	12.55	8.37
Required thickness of shells at the junction for pressure = P R/S	t	mm	10.87	7.25
Reinforcement is not required if delta > α, where delta = f(P/S)	P/S		7.2E-03	7.2E-3
Delta for large end = {11,15, 18, 21, 23, 25, 27, 28.5, 30} for [P/S = (1, 2, 3, 4, 5, 6, 7, 8, 9)/1000] (note 2)				
Delta for small end = {4,6,9,12.5,17.5, 24, 27, 30} for [P/S = (2, 5, 10, 20, 40, 80, 100, 125)/1000] (note 2)				
Angle delta Δ depends on P/S	Δ	deg	27.37	6.00
If $\Delta > \alpha$, no check is required for compensation. Else check for compensation				
Compensation check:				
Radial force per unit circ given in Eq. 5.2 = P R tan α/2	F^1	N/mm	433.0	288.7
Area required = $(1-\Delta/\alpha)$R F/S, if La < L, else Δ = 0	A_r	mm²	412.7	966.2
Limits of comp. from shells, (large = $2\sqrt{(R\ T)}$, small = $1.4\sqrt{(R\ T)}$	L	mm	268.3	178.9
Area available = If L_a < L, shell area is ineffective, cone area = $(T_C-t_C)\sqrt{(R\ T_C/\cos\alpha)}$				
Area at large end = $(T-t)\sqrt{(R\ T)} + (T_C-t_C)\sqrt{(R\ T_C/\cos\alpha)}$	A_a	mm²	377.3	
Area at small end = $0.78\sqrt{(R\ T)}[(T-t) + (T_C-t_C)/\cos\alpha]$	A_a	mm²		741.6

If $A_a < A_r$ additional area is added by providing the pad on the cone within a distance of *L* from the junction and centroid of the added area at a distance of 0.25L.

Because $A_a < A_r$ add 25 × 10 pad at 30 mm from the junction on the cone at both ends.

Note 1: If external axial force F_e and moment M are added, $F = [P\ R/2-Fe/(2\pi R) + M/(\pi R^2)]\tan\alpha$

Note 2: Reprinted from ASME 2019 BPVC, Section VIII-division 1, by permission of The American Society of Mechanical Engineers

For $\alpha > 60°$, the conical shell resembles a shallow shell and, finally, a circular plate and can be analyzed by assuming a (conservative) flat circular ring with both junctions as simply supported. For $30 < \alpha > 60°$, toriconical can be used or discontinuity stresses can be analyzed as per Table 7.1 or by Rules 4.3.11 and 12 of Ref. 2. It also gives the procedure to calculate the discontinuity stresses with knuckle.

A *toriconical reducer* consists of knuckle at the large end and flares at the small end to overcome the sharp discontinuity at the cylinder to cone junction and will greatly reduce discontinuity stresses.

Code rules for the toriconical reducer with $\alpha \leq 30°$: the thickness of the conical portion of the head is the same as the conical shell provided knuckle inside radius $\geq$ 6% of ID of cone and also $\geq$ three times thickness of knuckle. The knuckle portion thickness is the same as the torispherical head given by Eq. 5.14, L replaced by $R_i/\cos\alpha$.

5.4 TOROIDAL SHELL AND TUBE BEND

Notation: (see also Figure 5.2)

Points i = intrados, e = extrados
R_t = tangential radius = $R/\sin\theta$
R_o = mean bend radius (radius of revolution)
r and t = radius and thickness of torus (tube)

A toroid (torus) is the membrane shell developed by the rotation of the closed curve, usually a circle about an axis passing outside the generating curve as shown in Figure 5.2. While an entire toroidal shell, such as an automobile tube, is rarely used in pressure vessel components, segments are used for connecting two separate shells. Tube bend is the segment of the toroidal shell.

The principal membrane longitudinal stresses (σ_L) can be derived by static equilibrium between internal pressure P and membrane stress at any point b with radius R and angle θ from axis of rotation yy as shown in Figure 5.2.

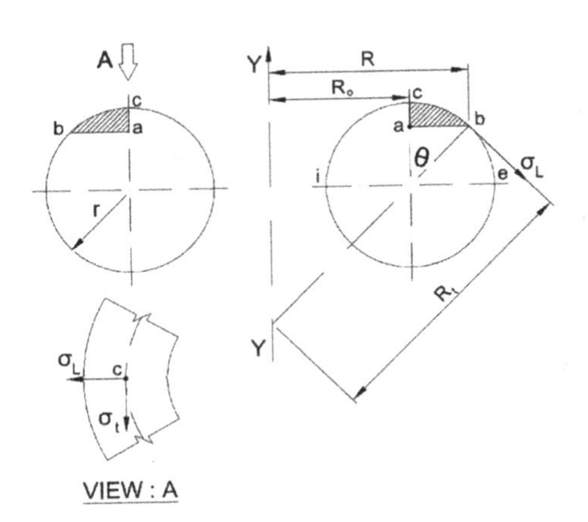

FIGURE 5.2 Toroidal shell.

The pressure effect on *ac* is equal and opposite and that on *ab* is resisted by the sin component of thickness over circumference with radius R at point b and given by the equation

$$P\pi(R^2 - R_o^2) = \sigma_L t \sin\theta(2\pi R)$$

Rewriting

$$\sigma_L = P(R - R_o)(R + R_o)/(2t R \sin\theta)$$

Substituting $R–R_o = ab = r\sin\theta$

$$\sigma_L = (Pr/2t)(R + R_o)/R \tag{5.3}$$

At intrados point i, where $R = R_o–r$

$$\sigma_L = (Pr/2t)[(2R_o - r)/(R_o - r)]$$

At extrados point e, where $R = R_o + r$

$$\sigma_L = (Pr/2t)[(2R_o + r)/(R_o + r)]$$

At point c, where $R = R_o$, σ_L is same as hoop stress of the cylinder with radius r, and is given by

$$\sigma_L = Pr/t$$

Longitudinal stress σ_L varies and is maximum at point c and minimum at e.

Tangential stress σ_t which is a circle with radius r is the same as longitudinal stress of the cylinder with radius r which is the same at any point and given by Eq. 5.4

$$\sigma_t = Pr/2t \tag{5.4}$$

This can also be derived from membrane theory Eq. 4.2

$$\sigma_t/R_t + \sigma_L/r = P/t$$
$$\sigma_t = P(R - R_O)/(2t\sin\theta) = Pr/(2t)$$

Applying the above theory for tube bends under internal pressure, stress σ_L at extrados (e) is less than the straight pipe and more at intrados (i) as per Eq. 5.3. Therefore, when the tube is bent, its thickness at the outer radius is reduced and the inner radius is increased, and this thinning can be allowed to the extent of the above equations. Some codes give this provision, but ASME codes do not allow. Actually when tested, the bend fails at the inner radius.

5.5 END CLOSURES

Shells require to be closed at both ends. Types of closures are flat, conical, and formed. Formed end closures can be many to the designer's expertise, but normally hemispherical, ellipsoidal, toriconical, and flat with knuckle (in order of reducing depth, cost, and resistance to pressure) are used. In addition to membrane and bending stresses in end closures, discontinuity stresses are induced in shells and end closures at the junction. Refer Chapter 7 for discontinuity stress analysis. The conical shell covered in 5.3 without the small end closes the connected cylindrical shell.

5.5.1 Flat

Flat is simple but requires higher thick and more weight compared to formed, practically economical only to small diameters and low pressures.

Refer Chapter 10 for flat plate fundamentals and analysis of flat circular and rectangular end closures.

5.5.2 Hemispherical

The hemi-spherical head is half of the spherical shell, and analysis is the same as that for the spherical shell in Section 5.2. However, discontinuity stresses exist at its junction with the shell due to the difference in hoop strain which is normally insignificant. Discontinuity stresses are covered in Chapter 7.

The code gives the same equation given to the spherical shell for stress but the limitation is different and given by $T \geq 0.356L$ or $P \geq 0.663S$, where L is the internal radius.

5.5.3 Ellipsoidal

Figure 5.3 shows the mean profile of the ellipsoidal head and notation.

The geometrical property of the ellipse is

$$x^2/a^2 + y^2/b^2 = 1$$

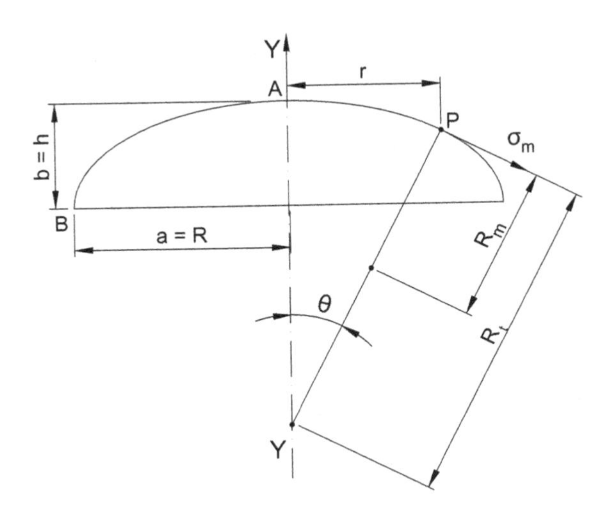

FIGURE 5.3 Semiellipsoidal head.

where

a and b are semiaxes of an ellipse.

a/b = R/h = k, and a, b, R, h, and other symbols are as shown in Figure 5.3. Practically, this relation is used with outside dimensions

Half of the ellipse has max radius in middle (crown) point *A* and is gradually reduced to min radius at 90° apart at point *B*. If k = 1, the depth *b* is equal to the radius of shell *a*, which is spherical. If k = 2, the depth is equal to half the radius of the shell and called 2:1 semiellipsoidal.

Radius R_t and R_m vary continuously and at any point *P* with latitudinal radius *r* (0 to a) and can be derived from the geometrical property and are given by

$$R_t = \sqrt{R^2(R/h)^2 + (1 - R^2/h^2)r^2} \tag{5.5}$$

$$R_m = R_t^3\, h^2/R^4 \tag{5.6}$$

Calculation of σ_m and σ_t: By equilibrium at latitudinal plane at *P*

$$2\pi r\, t\, \sigma_m \sin\theta = \pi r^2 P$$

Rewriting

$$\sigma_m = \frac{P\,r}{2t\sin\theta}$$

Substituting R_t for r/sin θ, we obtain

$$\sigma_m = \frac{P\,R_t}{2t} \tag{5.7}$$

Using Eq. 4.2 of membrane theory and substituting σ_m from Eq. 5.7 in Eq. 4.2, σ_t can be derived

$$\sigma_m/R_m + \sigma_t/R_t = P/t$$

$$\sigma_t = (P/t - \sigma_m/R_m)R_t = \left(P/t - \frac{P\,R_t}{2t\,R_m}\right)R_t$$

$$\sigma_t = \frac{P\,R_t}{t}\left[1 - \frac{R_t}{2R_m}\right] \tag{5.8}$$

At point A, $R_t = R_m$ and substituting in Eq. 5.6

$$R_t = R_m = R^2/h \tag{5.9}$$

Substituting R_t from Eq. 5.9 in Eqs. 5.7 and 5.8, and k for R/h

$$\sigma_m = \frac{PR^2}{2th} = \frac{kPR}{2t} \tag{5.10}$$

$$\sigma_t = \frac{P(R^2/h)}{t}\left[1 - \frac{1}{2}\right] = \frac{PR^2}{2th} = \frac{kPR}{2t} \text{ same as Eq. 5.10}$$

At point B, $R_m = h^2/R$ and $R_t = R$ substituting in Eqs. 5.7 and 5.8

$$\sigma_m = PR/2t$$

$$\sigma_t = (PR/t)\left[1 - \left(\frac{R^2}{2h^2}\right)\right] = (PR/t)\left[1 - k^2/2\right)] = \left(P\frac{R}{2t}\right)(2 - k^2) \tag{5.11}$$

If $R^2/2h^2 > 1$ or $k > 1.414$, σ_t will become compressive.

Hence as the k (>1) value (hemispherical = 1) increases, stress increases at A and decreases at B until k = $\sqrt{2}$, then reverses to compression and increases. At $k = 2$, max stress is the same at A and B (compressive at B) and given by Eq. 5.12 which is the same as the connected shell.

$$\sigma_t = PR/t \tag{5.12}$$

For $k > 2$ (code covers up to 3), stress at B increases sharply, and being compressive is not recommended. Due to discontinuity, moment is induced at the B shell to head joint, and the additional bending stress is induced. The elliptical profile undergoes deformation due to pressure and tries to become more spherical. Such deformation induces moment which will cause compression in the head at the joint and tension in the shell. Figure 5.4 shows the exaggerated deformation mean profile of the elliptical head in the dotted line. The shell at the tan point bends and tan point B will shift to C.

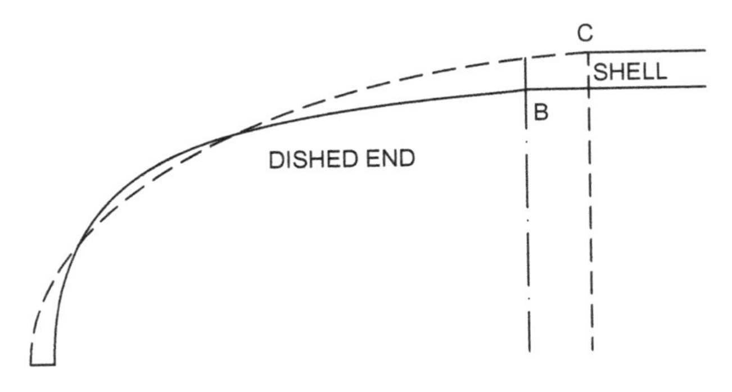

FIGURE 5.4 Deformation of the elliptical head.

The code[1] gives the equation for thickness with limitation $t/L \geq 0.002$

$$t = \frac{PDK}{2(S - 0.1P)}$$

where $K = (2 + k^2)/6$

For 2:1 semiellipsoidal ($k = D/2h = 2$)

$t = P\,R_i/(S–0.1P)$ or stress $= P(R_i/t + 0.1)$ appx $= P\,R_{mean}/t$, which is the same as Eq. 5.12.

For $0.0005 < t/L < 0.002$, refer section 5.5.5.

5.5.4 Torispherical

Figure 5.5 shows the torispherical mean profile and notation.Torispherical has two radii, smaller radius (r) at the end called knuckle and larger at rest of the center called crown radius (L). For a given k (≥ 1) $= R/h = a/b$ = shell radius to head depth, one extreme case is with the crown radius is infinity (straight line) and knuckle radius maximum ($= b$ flat) to other extreme with knuckle radius almost zero and crown radius minimum equal to spherical, an infinity number of knuckle and crown radii combinations exist. It can be noticed that the crown radius increases from its minimum value a to infinity and knuckle radius from zero to maximum b, and the ratio of the knuckle radius to crown radius goes from zero to maximum and back to zero. When this ratio is maximum, the resulting torispherical head most closely approximates to the elliptical head. The discontinuity increases from min at the first extreme case to max at other, and the junction of knuckle and crown produces additional discontinuity. Deformation of crown due to pressure is trying to become elliptical, that is, the radius reduces from A to B max at B.

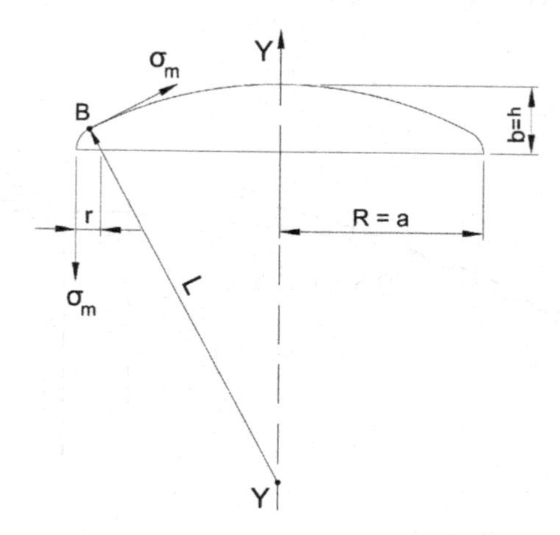

FIGURE 5.5 Torispherical head.

The maximum ratio r/L is related to k (2 to 3) by Eq. 5.13.

$$\frac{r}{L} = \frac{\sqrt{(k^2+1)} - k}{(\sqrt{k^2+1} - 1} \tag{5.13}$$

The membrane stress is maximum in the crown portion. As per the beam theory, the crown to knuckle junction is considered as a fixed boundary and the crown portion as the curved beam is under uniformly distributed load (UDL) due to internal pressure. In the knuckle portion apart from membrane stress, bending moment exists in the entire knuckle portion and max at the junction. Combined stress is max at the junction. Membrane stresses in crown portion as per membrane theory are

$$\sigma_m = \sigma_t = P\ L/2t \tag{5.14}$$

and in the knuckle portion at point B, the knuckle to crown junction is

$$\sigma_m = P\,r/2t$$

σ_t can be derived using Eq. 4.2 of membrane theory

$$\sigma_t / r_t + \sigma_m / r = P/t$$

Substituting σ_m from Eq. 5.14

$$\sigma_t = \frac{P\,r_t}{2t}$$

where

r_t = tangential radius of knuckle at the junction

The knuckle radius less than three times thickness and less than 6% of the inside diameter, and crown radius more than the inside diameter will have large local stresses due to discontinuity. The torispherical dished heads with the above limits are not permitted by the code. L and r are used for the inside radius in the code.

The code gives the equation of thickness for $t/L \geq 0.002$ and knuckle radius $r \geq 0.06L$ and $d\ L \leq Do$ as follows:

$$t = \frac{0.8885PL}{S - 0.1P}$$

If the limits except $t/L \geq 0.002$ exceed, the thickness is given as follows:

$$t = \frac{PLM}{2(S - 0.1P)} \tag{5.15}$$

where M is the stress intensification factor = $[3 + \sqrt{(L/r)}]/4$

For $0.0005 < t/L < 0.002$, refer section 5.5.5 below.

5.5.5 Ellipsoidal and Torispherical Heads with $0.0005 < t/L < 0.002$

Deformation is appreciable leading to buckling in the knuckle portion. The code gives equations to calculate min pressure (P_r) at which knuckle will fail by buckling [Appendix 1.4(f)(1), (2)] and is given by Eq. 5.16

$$\begin{aligned} &\text{If } P_e/P_y \le 1,\ Pr = 0.6P_e; \\ &\text{else if } P_e/P_y \le 8.29\ P_r = 0.408P_y + 0.192P_e; \\ &\text{else if } P_e/P_y > 8.29\ P_r = 2P_y \end{aligned} \tag{5.16}$$

where

P_e = internal pressure expected to produce elastic buckling = C_3 S_e

P_y = pressure expected to result in yield stress at the point of max stress = C_3 S_y

S_y and E = yield stress and elastic modulus at temperature and S_e = elastic buckling stress = C_1 E (t/r)

Constants a, b, ø, β, c, and R_e; a = D/2–r, b = L–r, $ø = \sqrt{Lt}/r$, β = Acos(a/b), and D = inside diameter of the head

L = crown inside radius for ellipsoidal = K_1 D

r = knuckle inside radius for ellipsoidal = K_2 D

K_2 = {0.5, 0.37, 0.29, 0.24, 0.2, 0.17, 0.15, 0.13, 0.12, 0.11, 0.1}

K_1 = {0.5, 0.57, 0.65, 0.73, 0.81, 0.9, 0.99, 1.08, 1.18, 1.27, 1.36}

For D/2h = {1, 1.2, 1.4, 1.6, 1.8, 2, 2.2, 2.4, 2.6, 2.8, 3}

K_1 = 0.44k + 0.02 & K_2 = 0.5/k–0.08, (k = D/2h)

c: [if ø ≥ β, c = a, else c = a/cos (β–ø)], R_e = c + r

Coefficients C_1 and C_2 = A + B r/D, where A and B are constants given below

For C_1: if r/D > 0.08, B = 0.692, and A = 0.605, else B = 9.31, and A = –0.086

For C_2: if r/D > 0.08, A = 1.46, and B = –2.6, else A = 1.25, and B = 0

$C_3 = t/[C_2R_e(0.5R_e/r-1)]$

5.6 FAILURE ANALYSIS

The stress equation of any part for any load will indicate the factors responsible for the increase of stress. For internal pressure load, the failure of the cylindrical shell is due to hoop stress. The increase of diameter and decrease of thickness will increase hoop stress in the cylindrical shell. When hoop stress reaches the yield point, yielding starts longitudinally at a point generally remote from support planes where the diameter is more due to ovality or reduced thickness due to erosion or corrosion. Due to yield, the diameter increases and thickness reduces, which will further increase

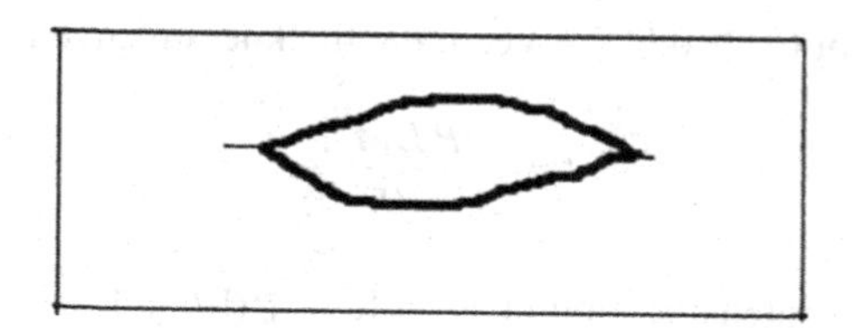

FIGURE 5.6 Hoop stress failure of the tube.

hoop stress even when the pressure is same, further yield, and so on. When stress reaches ultimate tensile stress and the shell opens longitudinally (bursts) with a small length and the fluid comes out with velocity, simultaneously pressure and stress start reducing and the opening length increases until the stress drops to yield stress. The shell looks like as shown in Figure 5.6.

Stresses due to other loads normally do not combine with hoop stress, but they do combine with longitudinal stress due to pressure. This combined stress is membrane cum bending and yields when stress reaches the yield point. Unlike in hoop stress, the progressive damage generally does not take place as long as the loads remain the same. When other loads increase, combined stress enters the plastic range and depending on the type of other load stress may relax for self-limiting or bending tensile failure may occur for primary bending loads. In the former case, brittle failure takes place on reaching fatigue cycles. In the latter case on tensile bending, failure takes place. On reaching yield across the entire tensile half area, yielding starts and on reaching ultimate tensile stress (UTS) circumferential crack develops at the extreme outer edge. As the fluid leaks out, pressure and stress reduce and crack development stops when stress drops to its yield point. As such materials used in pressure parts are ductile (long gap between yield point and UTS), the failure is time taking if not rare and less dangerous unlike hoop stress failure.

REFERENCES

1. Code: ASME S VIII D-1, 2019
2. ASME S VIII D-2, 2019

6 External Pressure

The action of external pressure on all pressure parts other than a flat plate is similar to the distortion of compressive flange of the I-beam in beam theory. The boundary planes are formed depending on the geometry of the part, and the portion under this boundary will buckle forming dish in reverse side with the mean surface area remaining the same. The max pressure the part can withstand depends on its geometry (D/t and L/D) and physical properties of the material (E, yield stress, and tangent modulus) at the component temperature. Higher the *d*/*t* and *L*/*d*, lower is the resistance to pressure. Basics of analysis are similar for all shells and are covered for cylindrical shells. For other shells, only *code*[1] rules are explained.

6.1 CYLINDER

Notation:

- P_C = theoretical critical external pressure at which a long cylindrical shell will collapse by buckling
- A = Strain coincident to P_c (same as in Ref. 2)
- B = allowable stress for external pressure depends on yield stress and *A*
- D = outside diameter of shell
- E = elastic modulus
- k = collapse coefficient
- f = critical buckling stress at P_c
- I = MI = moment of inertia
- L = length (effective) between two sections with resistance to retain the circularity against external pressure.
- L_C = critical length over which *A* remains constant.
- P_a = allowed pressure
- t = thickness
- v = Poisson's ratio

External pressure will induce hoop stress of the same value as internal pressure, but compressive instead of tensile. However, stress being compressive allowable stress is less than tensile and depends on *L*/*D* and *D*/*t* ratios.

6.1.1 Effects on Analysis for Variations in *L*/*D* and *D*/*t* Ratios

a. *A very short cylindrical shell* with $L < t$ or [$L/D < 0.25$ and $D/t < 4$] is not stressed, and entire pressure load is resisted by ends. It is rigid and transfers the load to both ends without any deformation. Therefore, stress induced can be neglected.
b. *A short and thick shell* ($L/D < 2.2$ and $D/t < 4$) fails by plastic yielding alone at high stresses close to the yield strength. The pressure effect is almost pure compression. Local buckling is not present. Stress and allowable stress are the same as internal pressure.

DOI: 10.1201/9781003091806-6

c. *Long and thin shell* ($L/D > 2.2$ and/or $D/t > 10$) buckles at stresses below the yield point. Corresponding critical pressure P_c is a function of D/t, L/D and E. For $L < L_c$, the shell buckles into multiple lobes and for $L > L_c$, P_c is independent of L, and the shell collapses into one or two lobes by buckling.

6.1.2 Critical Length

The critical length depends on D/t and is given by

$$L_c = 4\pi\sqrt{6/27}(1-\nu^2)^{0.25} D\sqrt{D/t}$$

Substituting $\nu = 0.3$ for steels

$$L_c = 1.11D\sqrt{D/t} \tag{6.1}$$

L_C is independent of pressure and is useful in selecting the spacing of stiffeners. Stiffeners are effective only if they are spaced at less than L_C. Tubes normally used are with $D/t < 25$ and $L_C > 5.55D$. Hence for $D/t < 25$, stiffeners are not required for tubes irrespective of their length

6.1.3 Critical Pressure

Critical pressure (P_c) can be derived by elastic theory for a strip (dS) subtend (dθ) at axis of unit length of cylindrical shell of radius (R_O) diameter (D) and unit width of circumference as follows using the flexure formula and is briefly explained below. For a complete derivation, refer 8.1 of Ref. 3.

On applying external pressure P, it will deflect by w and the radius increases to r, and $d\theta$ will increase to $d\theta + \Delta d\theta$, ds to $ds + \Delta ds$ as shown in Figure 6.1. Flexure formula Eq. 3.11 for beam, M = EI/R can be converted for the above strip as

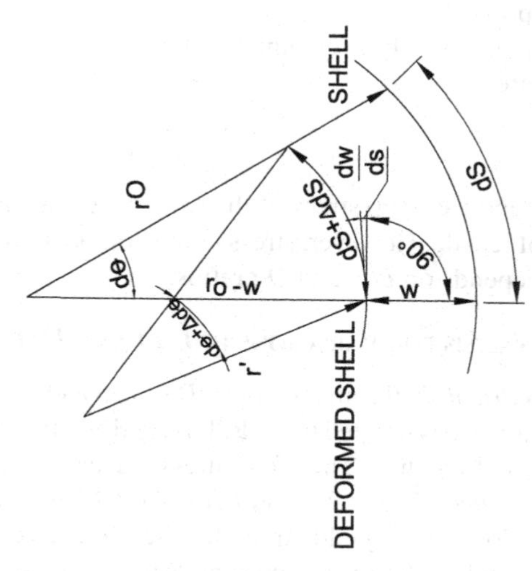

FIGURE 6.1 Cyl. shell deformation under ext. pressure.

$$M = EI\left(\frac{1}{r_o} - \frac{1}{r}\right)$$

Expressions for r, r_O, and $d\theta$ can be computed by geometry as

$$r_O = ds/d\theta,\ r = (ds + \Delta ds)/(d\theta + \Delta d\theta),\ \text{and}\ d\theta = (ds + \Delta ds)/(r_O - w)$$

Using the above equations, P_C can be derived by integration and is given by

$$P_C = 3EI/r_o^3 = 24EI/D^3 \tag{6.2}$$

Substituting $I = t^3/12$

$$P_c = \frac{2E}{(D/t)^3} \tag{6.3}$$

The adjacent metal either side of strip offers restraint to the longitudinal deformation of the strip. To allow this restraint, multiply Eq. 6.3 by $k/2$ to obtain Eq. 6.4

$$P_c = \frac{KE}{(D/t)^3} \tag{6.4}$$

Case 1: $L > Lc$ and $D/t > 10$
K is independent of geometry = $2/(1-v^2)$
For steels with $v = 0.3$, $K = 2.2$
Substituting the K value in Eq. 6.4

$$P_c = \frac{2.2E}{(D/t)^3} \tag{6.5}$$

Case 2: $L < Lc$ and $D/t > 10$
K is different from *case 1* and depends on both the ratios D/t and L/D and can be read from Figure 8.3 in Ref. 3.
The approximate equation for K is given by

$$K = 2.6\frac{\sqrt{(D/t)}}{L/D} \tag{6.6}$$

Substituting K from Eq. 6.6 in Eq. 6.4

$$Pc = \frac{2.6E}{(D/t)^{2.5}(L/D)} \tag{6.7}$$

P_a can be calculated by dividing P_c by the factor of safety (4 to 5) which depends on L/D, D/t, and tangent modulus.

6.1.4 Strain Coincident to P_c

Strain A is derived by substituting P_c from Eq. 6.4 for P in the hoop stress equation ($\sigma_c = P\,D/2t$) and stress f for σ_c

$$f = (k/2)E(t/D)^2 \tag{6.8}$$

Substituting f from Eq. 6.8 in equation $A = f/E$

$$A = (k/2)(t/D)^2 \tag{6.9}$$

6.1.5 Comparison of the above Equations with Code Equations

The value of A by Eq. 6.9 is the same as A in Ref. 2 for $D/t > 10$ and almost the same for $D/t < 10$. It may be observed from graph A of Ref. 2 that the A value is constant beyond a certain ratio of L/D. This is the same as case 1. Rest of the curve is the same as case 2.

The values of P_c and f depend on E, which is not valid above the proportional limit; instead tangent modulus (E_t) is relevant.

Et is much less than *E*. It is equal to stress/strain above the proportional limit and not constant. Due to the effect of tangent modulus, Eqs. 6.5 and 6.7 will deviate from actual values. Exact (practical) values can be calculated as per Ref. 2 rules.

Table 6.1 illustrates comparison of analysis of the above theory and code rules with examples.

TABLE 6.1
Comparison of Code and Theory 6.1 for Cylindrical Shells Under External Pressure

L-length	88	176	1600	1000	600	2000	3000	60000
D-OD	40	80	80	100	200	200	3000	3000
t-thickness	10	10	10	10	10	10	10	10
L/D	2.2	2.2	20	10	3	10	1	20
D/t	4	8	8	10	20	20	300	300
L_C/L	L< L_C	L< L_C	L> L_C	L> L_C	L< L_C	L> L_C	L< L_C	L> L_C
A.code	0.0959	0.0262	0.0174	0.0111	0.0045	0.0028	2.54E-04	1.30E-05
A = Eq. 6.9	0.0739	0.0261	0.0172	0.011	4.84E-03	2.75E-03	2.50E-04	1.22E-05
B code	118	118	105	101	88	82	62	1.2233
k_1	0.4585	0.1876	0.1876	0.1334	0.0251	0.0251	−0.0761	−0.0761
k_2	0.3333	0.1667	0.1667	0.1333	0.0667	0.0667	0.0044	0.0044
P_a code	54.0971	22.1339	19.6954	13.4734	5.8667	5.4667	0.2756	0.0054
L_C = Eq. 6.1	88.8	251.2	251.2	351.0	992.8	992.8	57677.3	57677.3
K	2.36	3.34	2.20	2.20	3.88	2.20	45.03	2.20
Pc	NA	1228.7	808.7	414.0	91.2	51.8	0.3139	0.0153

Equations from 6.1
A = K/[2(D/t)²]
Lc = 1.11*D√(D/t)
K = IF(L>Lc,2.2,2.6√(D/t)/(L/D)
Pc = KE/(D/t)³
Units: N, mm

Code equations
Pa = B[if(D/t>10, K2, K1)]
k_1 = 2.167/(D/t)–0.0833
k_2 = 4/(3B)
A code = if[D/t<4, 1.1/(D/t)², graph]
B = graph, AE/2 if A is left of graph

Temp = 260°C
E = 188200 MPa

B = AE/2

When the value of A or t/D is more than a certain value, the failure is by elastic–plastic buckling rather than by elastic failure. The value of A or t/D to which this effect is relevant depends on temperature, material, and shell geometry. For C&LAS with yield stress $Y > 207$ at 260°C, the value of $A = 0.00065$, above which the allowable stress B will not increase at the same rate but almost remain constant at higher values of $A > 0.03$.

Calculation of allowed pressure as per the code[1]

Case 1: For $D/t \geq 10$, $P_a = (4/3)B(D/t)$

Read A from graph or table in Ref. 2 for L/D (0.05–50) & D/t (4–1000).

Read B is approximate allowed stress and available from the curve or table in Ref. 2.

$B = A\ E/2$ If the B value is left of the curves

Case 2: For $D/t < 10$, Pa is calculated as follows:

If $D/t < 4$, A is calculated from Eq. 6.9 with $k = 1.1$

$$P_a = \text{smaller of } P_{a1} \text{ or } P_{a2} \tag{6.10}$$

where

$$P_{a1} = B\left[\frac{2.167}{D/t} - 0.0833\right]$$

$$P_{a2} = \frac{2S_b}{(D/t)}(1 - t/D)$$

$$S_b = Min\ (2S, 0.9\ yield)$$

P_a allowed pressure calculated from the code may be compared with Eq. 6.7 using the factor of safety in example Table 6.1.

Refer Appendix 3 in Ref. 2 for basis of establishing charts for A and B. The tangent modulus is used for buckling stresses above the proportional limit.

6.1.6 Design of a Circumferential Stiffener Ring

Because thickness to resist external pressure increases with effective length (L_e), stiffeners are provided to reduce L. Stiffeners can be given inside or outside and of any section. The stiffener ring will resist external load for a length (L) equal to sum of half the length on either side of stiffener. The pressure per unit length on the ring at collapse is equal to the product of P_t and L, which is equal to P_c in Eq. 6.2. Therefore,

$$P_t L = 24EI/D^3$$

where P_t = theoretical collapse pressure per unit length

I = MI per unit length

For the stiffener, P_t and I are considered as external pressure P and Is = MI of the stiffener ring required; the above equation can be written as $P\ L = 24E\ I_S/D^3$.

Rewriting, moment of inertia of ring I_S required is given by Eq. 6.11

$$I_s = \frac{P D^3 L}{24E} \tag{6.11}$$

Using the hoop stress equation ($f = \frac{PD}{2t}$), $P = 2f\,t/D$ and $E = f/A$

$$P/E = \left(\frac{2f\,t}{D}\right) / \left(\frac{f}{A}\right) = 2t\,A/D$$

Substituting, $2t\,A/D$ for P/E in Eq. 6.11

$$I_s = D^2 L\,t\frac{A}{12} \tag{6.12}$$

Moment of inertia of the ring and shell will act together to resist collapse. This combined M.I. required (I_C) may be considered as equivalent to that of a thicker shell with thickness (t_Y) given by

$$ty = t + As\ a/L$$

where A_S = cross-sectional area of the stiffener ring

Substituting t_Y for t in Eq. 6.12

$$I_C = D^2 L\left(t + \frac{A_S}{L}\right)\frac{A}{12} \tag{6.13}$$

Code[1] equations for comparison are

$$I_C = D^2 L(t + A_S/L)A/10.9$$

$$I = D^2 L(t + A_S/L)A/14$$

which are almost the same as Eqs. 6.12 and 6.13

For allowed pressure (B), the code gives equation

$$B = 3/4[PD/(t + A_S/L)$$

Strain A coincident to B can be read from the graph or table in Ref. 2 for the value of B.

For compiling the effective length of the shell for calculating provided MI (I_C), expression $1.1\sqrt{(D\,t)}$ is used.

Design philosophy of stiffeners: Because required combined MI of the stiffener plus shell is about 28% higher than that of the stiffener alone, and the actual combined MI is couple of times more than that of the stiffener alone; select a suitable bar stiffener to the required I_S which is simple. Select thickness of bar $\leq$ shell thickness, and depth $\leq$ 10 times thickness. In the case of depth limitation, use a channel with required I_S.

6.2 HEMISPHERICAL HEAD

Theoretical collapsing external pressure over the spherical shell of outside radius R is derived similar to the cylindrical shell in 6.1.2.2

$$P_c = 2E/\left\{(R/t)^2\sqrt{[3(1-v^2)]}\right\}$$

with v = 0.3 for steels,

$$P_c = 1.21E(t/R)^2 \tag{6.14}$$

Applying the factor of safety P_a can be calculated.

The code[1] gives equations for calculating P_a below.

Strain at critical pressure $A = 0.125(t/R)$

Read max allowable stress B for A and yield stress from the graph or table in Ref. 2.

Allowable pressure is given by the equation combining B and hoop stress and given by

$$B = PR/t$$

Substituting P_a for P

$$P_a = B(t/R)$$

If A is to left of the curve $P_a = 0.0625E(t/R)^2$

The difference of code equation and Eq. 6.14 is due to tangent modulus used in code.

6.3 ELLIPSOIDAL AND TORISPHERICAL HEADS

P_a is min at crown point and more in the knuckle region unless buckling governs. P_c is the same as Eq. 6.14 with R as the crown radius for torispherical and k D (D is outside diameter) as the radius for ellipse.

The code[1] allows two procedures.

First: Allowed pressure on the convex side is 1.67 times that of pressure on the concave side.

Second: The same procedure as the spherical head except crown outside radius in place of R. In the case of elliptical head whose radius varies from maximum at center to minimum at end, the equivalent radius $K\ D$ depends on $D/2h$, where h is the outside height of dish and d is the diameter of the head.

For 2:1 semiellipsoidal $D/2h = 2$, $k = 0.9$, for other values of $D/2h$, k is given below.

k = {0.5, 0.57, 0.65, 0.73, 0.81, 0.9, 0.99, 1.08, 1.18, 1.27, 1.36}

For {D/2h = (1, 1.2, 1.4, 1.6, 1.8, 2, 2.2, 2.4, 2.6, 2.8, 3)}

6.4 CONICAL PARTS

For half cone angle $\alpha \leq 60°$, analysis is the same as the equivalent cylinder with $t_e = t\ cos\alpha$, and L_e equivalent length (as defined in 6.1) = $(L/2)[1 + (R_L/R_s)]$, and $L/2$ for the head (Figure UG 33.1 of the code[1] shows the details of *Le*). For $\alpha > 60°$, the cone is analyzed as a flat plate with diameter equal to the large diameter of the cone. An *eccentric cone* is analyzed with large or small α.

As described in 5.3 and Figure 5.1, an additional force component in vertical direction (F = P R tan α/2) induces discontinuity stresses for external pressure also. Forces *F*, *T*, and *N* in case of external pressure are opposite to those shown in Figure 5.1. The stresses are compressive at the small end and also compressive if $F < P R$, that is, if α is more than about 63°. The compensation procedure is similar to that described in calculations in Table 5.1. The difference is mainly the area required. The analysis is illustrated by calculations in Table 6.2. Because the stresses are

TABLE 6.2

Compensation calculations due to external pressure at the cone to shell junction

Data: P = 0.5, S = 138, E = 190000 in Mpa, units: mm, N UOS		**Large.L**	**Small.s**
Outside Radius of cone, material SA-516 70, temp = 234°	Rc	760	510
Thickness of cone (half cone angle α=39°)	Tc	10	10
Thickness of cylindrical shells	T	10	8
Outside radius of the shell, material SA-516 70	R	760	508
Length of the cylindrical shell	L	300	160
Calculations:			
Required thickness of the cone for pressure as per code	tc	8	5.8
Required thickness of shells for pressure as per code	t	8	5.8
Reinforcement is not required for large end if α < Δ, where Δ = f(P/S), not applicable to small end.	P/S	0.00362	
S*E/(Sr*Er) for the large end, E and S are the same for all parts	k	1	
Angle in degrees for large end depends on P/S given below	Δ	6.082	
Δ = {5, 7, 10, 15, 21, 29, 33, 37, 40, 47, 52, 57, 60} for			
P/S = (0.2, 0.5,1,2, 4, 8, 10, 12.5, 15, 20, 25, 30, 35)/100, (note 1)			
Compensation check: Fe and Me are external axial force and moment and zero			
Limits of compensation for shells, large = 2√(R T), and small=1.4√(R T)	h	174.4	89.2
Force/unit circ = P R/2 + Fe/(2πR) + Me/(πR²), N/mm	F	190	127
because L > h, Ar = Rc k F tanα/S	Ar	814.3	379
Aa = Area available in shell and cone: if L < h, shell area is ineffective and Aa = Eq.1 for large end and Eq. 2 for small end, else Aa = Eq. 3 for large end and Eq. 4 for small end.			
Eq. 1 = 0.55(Tc/cosα)√(2Rc Tc), Eq. 2 = 0.55√(2Rc Tc)(Tc - tc)/cosα			
Eq. 3 = 0.55(T+Tc/cosα)√(2Rc T), Eq. 4 = 0.55[(T-t)+(Tc-tc)/cosα]√(2R T)			
since L > h, eq. (3) & (4) are applicable	Aa	1555	378

Aa < Ar for small end. Reinforcing ring external bar 50 x10 is provided.

Note 1: Reprinted from ASME 2019 BPVC, Section VIII-Division 1, by permission of The American Society of Mechanical Engineers.

compressive for external pressure, the required compensation is provided by a ring instead of pad.

Reinforcement ring moment of inertia calculation for small end: units N, mm, MPa

L_C = 311 = Length of cone = $(R_L - R_s)/\tan\alpha$

L_e = 400 = Length of the cone between stiffeners, max = $\sqrt{L_C^2 + (R_L - R_S)^2}$

A_S = 50*10 = 500 = Area of the stiffener ring

A_t = 3140 = Effective area of (shell + cone) + A_s = $(L\,T + L_e\,T_c)/2 + A_s$

M = 79.27 = $L_c/2 - R\cdot\tan\alpha/2 + (R_L^2 - R_s^2)/(C\,R\tan\alpha)$, C = 6 (3 for the large end)

f = 39.64 = $P\,M + F_e\tan\alpha$, if < 0, NA and FEA may be carried out. $F_e = 0$

B = 9.66 = Allowable stress = $1.5f\,R_c/A_t$

A = 0.000104 = From curve (Ref. 2) depends on B and Y, if B is left of graph, A = 2B/E

I_S = 23801 = Required MI of the stiffener ring = $A\,4R_C^2\,A_t/14$

I_S = 30571 = Required combined MI of shell/cone/ring = $A\,4R_c^2\,A_t/10.9$

I = 104167 = MI of the stiffener ring provided

I = 319640 = Combined MI of shell/cone/ring provided [shell length ≤ $1.5\sqrt{RT}$]

Ring shall be placed on shell within a distance ($\sqrt{RT}$ = 64) and its centroid within distance ($0.25\sqrt{RT}$) from junction and on cone if L < h.

REFERENCES

1. Code: ASME S VIII D-1, 2019.
2. ASME S IID, 2019.
3. *Process Equipment Design*, L. E. Brownell & E. H. Young, 1959.

7 Discontinuity Stresses

Notation UOS: S: basic allowed stress, P: internal pressure, σ: stress, v: Poisson's ratio, E: elastic modulus, y: radial deflection, θ: rotation, R: mean radius, t: thickness, Q: edge shear force, and M: edge moment.

Suffixes: c: circumferential and L: longitudinal.

7.1 GENERAL

The pressure vessel consists of axially symmetrical elements of different geometries, thicknesses, and physical properties, and dissimilar materials with different expansion coefficients. Individual shell elements are allowed to expand (deform) freely as separate sections under any variation mentioned above; each such shell element would have an edge radial displacement and rotation of the meridian tangent that would differ from that of the adjacent shell element or component. Because both are to deflect and rotate together, the difference in such deflection and rotation results in local shell deformations and stresses to preserve the physical continuity of the shell. Stresses induced by such interaction of two shell elements at their junction are called discontinuity stresses. Some examples are junctions between the cylindrical shell and hemispherical, ellipsoidal, torispherical, and conical heads; crown to knuckle joint of the torispherical head, butt joint in shells of dissimilar thickness and properties, and shell to nozzle or other attachments.

Discontinuity stresses are usually not serious under static loads such as internal pressure with ductile materials if the degree of discontinuity is kept low by design, but they become important in cyclic loading. Under steady loading, they are self-limiting, meaning that a small plastic relaxation will reduce the acting force. The discontinuity stresses combine with longitudinal (meridian) stresses, and their effect in the circumferential direction is limited. However, the combined stress in the longitudinal direction will generally (except spherical shells) may not exceed allowed, because the longitudinal membrane stress due to pressure is less than half of that allowed. In average application, they will not lead to failure but fail after several cycles of operation. These stresses will become important factors in fatigue design where cyclic loading is a consideration. Design of the juncture of the two parts is a major consideration in reducing these stresses. Codes specify design rules in restricting the degree of discontinuity. Few tips are listed in 7.9.

Discontinuity stresses at the junction of cylindrical shells with other shells and heads under internal pressure load are covered in this chapter, and discontinuity stresses at the junction of shells with attachments such as nozzles, supports, etc., due to pressure and external forces are covered in Chapter 8.

DOI: 10.1201/9781003091806-7

7.2 GENERAL PROCEDURE

Step 1: *Radial deflection y due to internal pressure*: Calculate local radial deflection at discontinuity of both adjacent components as free parts (not connected) due to pressure load as given below by membrane and elastic theories. Rotation is zero as membrane shells do not rotate due to pressure. Radial deflection is equal to radial strain multiplied by radius.

Using equations: Radial strain for cyl. shell = σ_c/E-v σ_L/E, σ_c = P R/t, and σ_L = P R/2t

$$y = P\,R^2(1 - v/2)/(E\,t)$$

Or *y* per unit *P*

$$y/P = R^2(1 - v/2)/(E\,t) \tag{7.1}$$

Similarly, *y* per unit *P* can be derived for other shells and heads.

Step 2: *y and θ due to Q and M*: Due to the difference in *y* in edges of both shells under internal pressure, *edge shear forces* equal and opposite are induced. Due to *Q*, both edges deform and rotate. Rotation may be different for both, due to which equal and opposite *edge moments* are induced. *y* and *θ* due to *Q* and *M* above as well as due to external and thermal loads are added. The vector total *y* is equal and that of *θ* will be equal and opposite on both edges forming two equations, and by solving two unknowns, *Q* and *M* are obtained.

y and θ *due to Q* and *M* are calculated by elastic theory and can be expressed in matrix Eq. 7.2.

$$[y, \theta] = [K]\,[Q, M] \tag{7.2}$$

where K = flexibility matrix 2 × 2.

Sign convention: outward displacements are positive, and anticlockwise rotations are positive. To rotate together, signs of both rotations are opposite unlike displacements both of which have the same sign. Because it is not easy to feel the direction of movements without knowing the directions of *Q* and *M*, the direction of *Q* and *M* is assumed so that they are equal and opposite as shown in Figure 7.1. Even if they

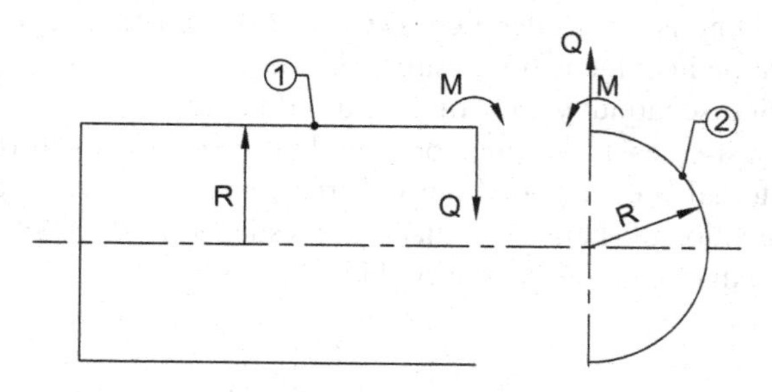

FIGURE 7.1 Cylinder to hemispherical head.

TABLE 7.1
Elements of [K] for Discontinuity Stresses

y/θ	P/Q/M	Cylinder	Sphere	Cone
y_p	P	$R(\sigma_c - v\,\sigma_L)/E$	$R(\sigma_c - v\,\sigma_L)/E$	$R(\sigma_c - v\,\sigma_L)/E \cos\alpha$
y_q/Q	Q	$1/(2\beta^3 D) = d$	$(2R^2\beta)/(E\ t)$	$d \cos^2\alpha/(k\ n)^2$
y_m/M	M	$1/(2\beta^2 D)$	$(2R^2\beta^2)/(E.t)$	$d\ \beta \cos\alpha/(k^2\ n^3)$
θ_q/Q	Q	$1/(2\beta^2 D)$	$(2R^2\beta^2)/(E.t)$	$d\ \beta^2 \cos\alpha/(k^2 n^3)$
θ_m/M	M	$1/(\beta\ D)$	$(4R^2\beta^3)/(E\ t)$	$2d\ \beta^2/(k\ n^3)$

are wrong, it will correct giving the negative sign in the calculated values of Q and M. In the calculation, correct signs as per the above convention shall be given for y & θ while forming Eq. 7.2

Refer Table 7.1 for elements of flexibility matrix [K] for cylindrical, spherical, and conical shells/heads for calculating deflection and rotation due to pressure, shear force, and moment loads, which are calculated as per membrane and elastic theory. The stiffness is valid for continuity of diameter and thickness for a length more than a limit called the bending boundary zone of the shells.

Step 3: *Calculation of radial deflection and rotation due to thermal loads*: Radial deflection due to temperature (T) is equal to $\alpha\ R\ T$, where α = coefficient of thermal expansion. No rotation is induced due to temperature. Generally, the temperature is the same for both and has no effect. In the case of dissimilar materials with a difference in α, radial deflection is effective. α for C&LAS is the same, but it is different for high alloys and stainless steels.

Step 4: *Calculation of Q and M*: Add y and θ due to all loads calculated above for both shells and form two equations with two unknowns *Q and M*, and by solving, values of *Q and M* are obtained.

Step 5: *Calculation of combined longitudinal and tangential stresses*: Combined longitudinal σ_{LC} and circumferential σ_{CC} (tangential stresses σ_{tC}) due to pressure and Q and M are given by Eqs. 7.3 and 7.4.

$$\sigma_{LC} = \sigma_p + 6M/t^2 \tag{7.3}$$

$$\sigma_{tC} = \sigma_p + \sigma_m + \sigma_b \tag{7.4}$$

where

σ_p = pressure stress (long or tangential)
σ_m = membrane stress due to Q & M = E × strain = $E(Q\ y_q + M\ y_m)/R$
σ_b = bending stress due to Poisson's ratio effect of M = $6v\ M/t^2$

Allowed stress for $\sigma_p + \sigma_m$ is 1.5 times basic allowed stress (S) as σ_m is local membrane stress and 3S for membrane + bending stress $\sigma_p + \sigma_m + \sigma_b$ as σ_b is secondary bending stress. The discontinuity stresses for cylindrical shell closures with formed heads are less compared to higher allowed stresses and are overlooked.

7.3 CYLINDRICAL TO HEMISPHERICAL HEAD

The discontinuity stresses at the cylindrical to hemispherical head junction for internal pressure and other loads can be calculated using the procedure given in *section* 7.2 and using equations in Table 7.1. Detailed calculations are illustrated by example in Table 7.2. Figure 7.1 shows the notation and forces at the junction. Discontinuity stresses at the junction are marginal. σ_{Lc} of both will increase marginally and σ_{cc} of the cylinder will reduce and reverse for the head as it resists the free movement of cylinders due to pressure. Therefore, cylinders at the joint are safer than away from the

TABLE 7.2
Calculation of Discontinuity Stresses at the Junction of the Cylindrical Shell to Hemispherical Head Under Internal Pressure

Defection = y, rotation = θ	Sym	Units	1-cyl	2-HShead
Mean radius	R	mm	2500	2500
Thickness	t	mm	10	5
Pressure	P	Mpa	0.5	0.5
Temperature	T	°C	130	130
Elastic modulus for mat. SA516 70 at T = 130°	E	MPa	200000	200000
Poisson's ratio for the material	v		0.3	0.3
Coefficient of thermal exp. for mat. at T	α		1.2E–07	1.2E–07
Local MI per unit circumference = $t^3/12$	I	mm^3	83.33	10.42
Shell constant = $[3(1-v^2)/(R\ t)^2]^{0.25}$	β	1/mm	8.13E-03	0.01150
Shell constant $E\ I(1-v^2)$	D	Nmm	1.52E + 07	1.90E + 06
y due to pressure P, θ = 0				
Circ. Stress (1) = P R/t, (2) = P R/2t	σ_c	Mpa	125	125
Long. stress (1) = P R/2t, (2) = P R/2t	σ_L	Mpa	62.5	125
Radial $y = R(\sigma_c - v\ \sigma_L)/E$	y_p	mm	1.33	1.09
y, θ due to Q - unit shear force per unit circumference				
Radial y in (1) = $-1/(2\beta^3 D)$, (2) = $(2R^2\beta)/(E\ t)$	y_q/Q	mm	–0.061	0.144
Radial θ in(1) = $-1/(2\beta^2 D)$, (2) = $-(2R^2\beta^2)/(E\ t)$	θ_q/Q	rad	–5E-04	–1.6E-03
y, θ due to M - unit moment per unit circumference				
Radial y in (1) = $-1/(2\beta^2 D)$, (2) = $-(2R^2\beta^2)/(E \cdot t)$	y_m/M	mm	–5.0E-04	–1.7E-03
Radial θ in (1) = $-1/(\beta\ D)$, (2) = $(4R^2\beta^3)/(E\ t)$	θ_m/M	rad	–8.1E-06	3.8E-05
y due to temperature, θ = 0				
Radial y = α R T, NA if temp. & mat. is same	y_t	mm	0.040	0.04
y, θ due to external forces	y_e, θ_e		0	0
Total $y = y_p + Q\ y_q/Q + M\ y_m/M + y_t + y_e$, and $y_1 = y_2$ – Eq. 7.1			1.230	1.230
Total $\theta = Q\ \theta_q/Q + M\ \theta_m/M + \theta e$, and $\theta_1 = \theta_2$ – Eq. 7.2			–9.3E-04	–9.3E-04
Solving Eqs. 7.1 and 7.2	Q	N/mm	1.330	1.330
	M	N	33.3	33.3
Combined longitudinal stress = $\sigma_L \pm 6M/t^2$	σ_{Lc}	MPa	64.5	133.0
Displacement due to Q & M = $Q\ y_q/Q + M\ y_m/M$	y_{qm}	mm	–0.098	0.136
Comb. circ. stress = $E\ y_{qm}/R + \sigma_c \pm 6v\ M/t^2$	σ_{cc}	MPa	117.7	138.3

joint, but head combined stress σ_{cc} may exceed basic allowable stress but is generally less than yield stress. The effect of this discontinuity due to pressure can be ignored.

7.4 CYLINDER TO OTHER END CLOSURES

7.4.1 Semiellipsoidal Head

Analysis of discontinuity stresses at the junction of the cylinder to semiellipsoidal head is very complicated due to continuous variation of radius. However, discontinuity stresses are not of concern as this stress is much less than at the center due to a smaller radius at tan point. Therefore, combined stress including discontinuity stresses is normally less than allowed. Max stress in 2:1 semiellipsoidal head (k = diameter/height = D/h = 2) at the junction is almost the same as shell hoop stress under internal pressure load. Therefore, it is popularly used. For a higher value of k, stresses are considerable, and these heads with $k > 2$ are generally not used.

7.4.2 Torispherical Head

There are two junctions of discontinuities, knuckle-crown and head-shell. Both are close together and affect each other. Total combined stresses in the knuckle region are several times higher than 2:1 semiellipsoidal. Torispherical heads are, therefore, not suitable for high pressures.

7.4.3 Flat Plate

The relevant element of flexibility matrix Eq. 7.2 for the flat plate to be used in discontinuity stresses at its joint with any shell can be computed from flat plate fundamentals as follows:

Ref. 1 Table 11.2 case 10a; rotation θ for the free plate under pressure is given by Eq. 7.5

$$\theta = \frac{P R^3}{8D(1+\nu)} \tag{7.5}$$

Case 10b, M-edge moment for fixed edge under pressure is given by Eq. 7.6,

$$M = P R^2/8 \tag{7.6}$$

By dividing Eq. 7.5 by 7.6, free radial rotation of edge of flat plate θ_m is given by

$$\theta_m = \frac{R}{D(1+\nu)}$$

Other three elements are very less and taken as zero. D is the plate constant the same as the shell constant and R outside radius of the plate.

For a rigid flat plate (encased in concrete or plate of large thickness), y and θ due to Q and M are zero.

Rest of the analysis is the same as section 7.2. Discontinuity stress in the flat plate = $M/Z = 6M/t^2$

7.5 CYLINDER TO CONE

Analysis of discontinuity stresses at the cyl–cone junction is as per 7.2 and similar to the cyl–spherical head described in Table 7.2. The difference is that the discontinuity is not due to the difference in radial displacement but due to cone inclination (α). Higher the inclination higher the discontinuity stresses, and for $\alpha<30°$, the discontinuity stresses are marginal and generally verified by the compensation method (Table 5.1).

Step 1: there will be no deformation at the cyl–cone junction due to pressure.

Step 2: y and θ due to Q and M.

The equations for cylinders are shown in Table 7.1 and used for example in Table 7.2. y and θ can be derived for the cone by the equivalent cylinder method. The equivalent cylinder replaces the cone on inclined axis with a larger radius cylinder than the connected cylinder as shown in Figure 7.2. The radius is obtained by dividing the connected cylinder radius by cosα. The force and displacements Q and y used are normal to the cylinder axis, which are components of the same normal to cone axis.

Equivalent cylinder method

Notation: also refer Figure 7.2 and Table 7.2

α = half cone angle

n = ratio = thickness of cone/thickness of the connecting cylinder

R and t = mean radius and thickness of the cylinder

D and β = shell constants of the cylinder defined in Table 7.2

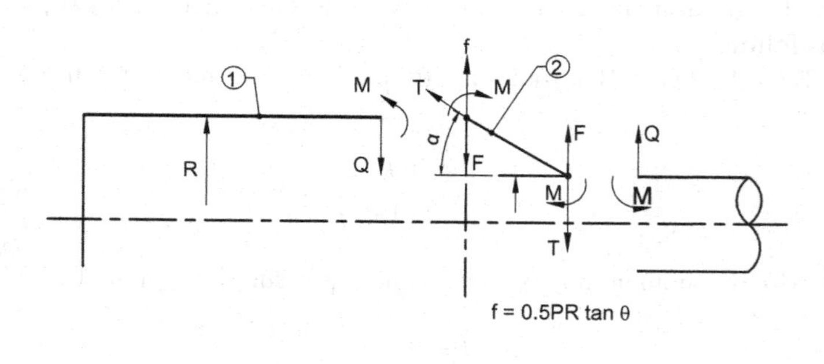

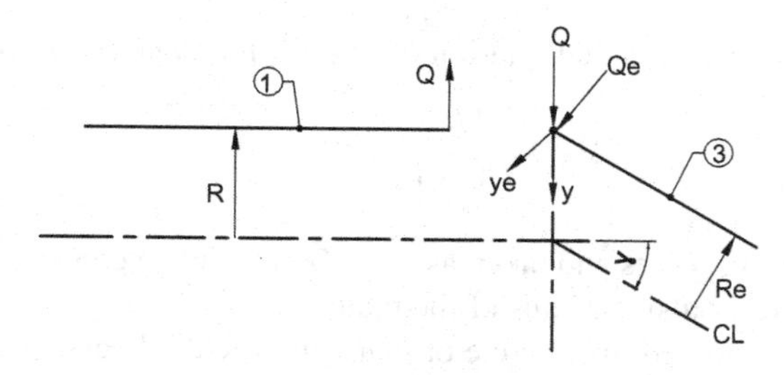

FIGURE 7.2 Cylinder to cone junction.

For condensing expressions below, displacement of the cylinder due to unit Q is denoted by d.

$$\frac{y_q}{Q} = \frac{1}{2\beta^3 D} = d \tag{7.7}$$

Dimensions of the equivalent cylinder of the cone

t_e = Thickness = n t

R_e = Mean radius = R/cos α

$$y_e = y/\cos\alpha \tag{7.8}$$

$$Q_e = Q\cos\alpha \tag{7.9}$$

$$D_e = EI(1-v^2) = E(1-v^2)t_e^3/12 = E(1-v^2)n^3t^3/12 = Dn^3 \tag{7.10}$$

$$\beta_e = \left[\frac{3(1-v^2)}{(R_e t_e)^2}\right]^{0.25} \tag{7.11}$$

Substituting (n t R/cos α) for $R_e\, t_e$ in Eq. 7.11

$$\beta_e = \left[\frac{3(1-v^2)}{(n\,t\,R/\cos\alpha)^2}\right]^{0.25} = \sqrt{(\cos\alpha/n)}\left[\frac{3(1-v^2)}{(R\,t)^2}\right]^{0.25} = k\beta \tag{7.12}$$

where

$$k = \sqrt{(\cos\alpha/n)}$$

y and θ are for pressure and Q and M of the equivalent cylinder of the cone are calculated by using the equations of the cylinder in Table 7.1 and equivalent parameters computed in Eqs. 7.7 to 7.12 as given below.

$$y_{qe} = \frac{Q_e}{2\beta_e^3 D_e}$$

Substituting k β for β_e, D n^3 for D_e and d for $1/(2\,\beta^3 D)$ from Eqs. 7.7, 7.10, and 7.12

$$\frac{y_{qe}}{Q_e} = \frac{1}{2k^3\beta^3 n^3 D} = \left[\frac{1}{2\beta^3 D}\right]\left[\frac{1}{(k\,n)^3}\right] = \frac{d}{(k\,n)^3}$$

Substituting y_q for y_{qe} and Q for Qe from notation

$$\frac{y_q}{Q} = \frac{d\cos^2\alpha}{(kn)^3} \quad (7.13)$$

Similarly, y_m/M, θ_q/Q, and θ_m/M can be derived and tabulated in Table 7.1

Step 3: refer step 3 of 7.2 and Table 7.2.

Step 4: calculation of Q, F, and M

Shear force will not be the same for the cone to cylinder like the cylinder to sphere at the discontinuity point due to the additional radial force P R $\tan\alpha/2$. Therefore, denote radial force F, deflection y_f, and rotation θ_f for the equivalent cylinder of the cone in the place of Q, y_q, and θ_q of the cylinder at the edge of cone. By equilibrium of forces at junction

$$Q + F = PR(\tan\alpha)/2 \quad (7.14)$$

Radial displacement due to pressure is almost the same for both and hence omitted, and rotation is assumed as the same although it is not theoretically true. By equating *y and θ* due to all loads at the shell–cone junction for the cylinder and eq. cylinder of cone, two more equations are formed as given below.

$$[Qy_q/Q + My_m/M]_{cyl} = [Fy_f/F + My_m/M]_{cone} \quad (7.15)$$

$$[Q\theta_q/(Q,F) + M\theta_m/M]_{cyl} = [Q\theta_f/F + M\theta_m/M]_{cone} \quad (7.16)$$

Unknowns (Q, F, and M) can be solved with these three equations Eqs. 7.14 to 7.16

Step 5: combined longitudinal and tangential stresses

Combined long and tangential stresses due to pressure and Q, F, and M can be calculated using membrane theory and the above equations and are given below.

Cylinder:

$$\sigma_L = \frac{PR}{2t} \pm \frac{6M}{t^2}$$

$$\sigma_t = E \times \text{strain due to Q \& M} + P\,R/t \pm 6v.M/t^2$$

$$\sigma_t = \frac{E(y_q Q + y_m M)}{R} + \sigma_c \pm 6vM/t^2$$

Cone:

$$\sigma_L = \frac{PR\cos\theta}{2t} \pm \frac{6M}{t^2}$$

$$\sigma_t = E \times \text{strain due to Q \& M} + P\,R/t\cos\alpha \pm 6v.M/t^2$$

$$\sigma_t = E(y_f F + y_m M)/R + \sigma_c \pm 6vM/t^2$$

7.6 ANALYSIS IN THE CYLINDER FOR EDGE FORCES AT ANY DISTANCE FROM THE EDGE

Deflections and rotations and shear force and moments at any distance (x) from the edge reduce as the distance increases and become insignificant at a distance *L*.

The following equations can be used to calculate for the cylindrical shell for Q and M (Ref. 2)

$$M_x = Q\,C\sin(\beta x)/\beta + M\,C[\cos(\beta x) + M\sin(\beta x)]$$

$$F_x = Q\,C[\cos(\beta x) - \sin(\beta x)] - 2M\,C\sin(\beta x)$$

$$y_x = \frac{Q\,C\cos(\beta\,x)}{2\beta^3 D} + \frac{M\,C[\cos(\beta\,x) - \sin(\beta\,x)]}{2\beta^2 D}$$

$$\theta_x = \frac{Q\,C[\cos(\beta\,x) + \sin(\beta\,x)]}{2\beta^2 D} + \frac{M\,C\cos(\beta\,x)}{\beta D}$$

where

$C = e^{-\beta x}$

L is the bending boundary zone and approximately $= 4\sqrt{Rt}$ for the cylindrical shell.

7.7 DESIGN TIPS

1. Different thicknesses at the junction will induce additional stresses and can be reduced by taper 1:3 minimum.
2. Eccentricity of the mean radius will induce additional stresses and can be reduced by matching the mean radius.

REFERENCES

1. Formulas for stress and strain, Raymond J. Roark and Warren C. Young, 5th edition
2. Pressure vessel design handbook, H.H. Bednar, 1986.

8 Local Stresses

The process requires several types of attachments to the shell such as reducers, expanders, branch pipes, end closures, bends, expansion bellows, supports, guides, stiffeners, flanges, local reinforcements, etc. Local stresses are induced due to discontinuity and stress concentration in shells due to the openings and above attachments under pressure and external loads.

For spherical shells, the effect is the same all around the opening. In cylindrical shells, the effect is maximum in the circumferential plane and reduces gradually as rotated and minimum at 90° in the longitudinal plane as the pressure stress is half that in the circumferential plane. In rectangular shells, the effect of openings is generally less due to negligible membrane stress compared to bending stress. Unless the thickness of the attachment nozzle local to the junction is sufficiently higher than that required for pressure load, the shell requires local reinforcement by a pad and/or margin in its thickness.

Supports other than saddles and skirts are generally over a limited portion of shell. The reaction loads are similar to external loads through attachments and covered in this chapter. Local stresses due to the reaction of saddle and skirt supports are covered separately in Chapter 11.

8.1 OPENINGS UNDER PRESSURE

Openings are mainly due to nozzles. The required resisting area in the cylindrical shell for pressure load in any plane (circumferential to longitudinal) will reduce due to openings, and membrane stress will increase locally at the shell to nozzle joint. In addition, bending stress is induced due to stress concentration and shell bulges longitudinally local to opening and cracks by fatigue after reaching its life cycles.

8.1.1 Analysis

Analysis is carried out by two methods: stress analysis and compensation.

(a) *Stress analysis method*: This method is carried out by calculating discontinuity (membrane and bending) stresses local to junction. Membrane (circumferential) stress is derived based on static equilibrium over a length of shell local to opening. Mathematical solution for bending moment can be derived by certain assumptions. It is complicated and involves errors due to assumptions. Specific finite element analyses are available one of which is Nozzlepro.

(b) *Compensation method*: The compensation method is based on the basics that the area required to resist pressure in the shell in any plane removed due to opening is to be compensated from the area available in the shell and nozzle in excess of the area required to resist pressure, within a limited distance from the opening and by

DOI: 10.1201/9781003091806-8

intentionally reinforcing (adding pad on shell) around the nozzle. This method has the following limitations.

1. It is not logical for small openings.
2. Conservative up to a certain size.
3. For sizes nearer to the shell diameter, the stresses may exceed there by the need of addition check by the stress analysis method.

The following factors are considered in design.

1. Width of pad > 16 times thickness may buckle.
2. Adding more pad area than required creates a too hard spot on the vessels, and large secondary stresses due to the shell restraint can be produced in consequence.
3. Providing a larger radius of corners and fillet welds will reduce stress concentration.

Openings are categorized as small, single, multiple, continuous, and large bore for the purpose of the compensation method. The compensation method for the above openings is explained in the following sections.

8.1.2 Small Openings

For isolated (single, defined in 8.1.3) openings up to certain size, no analysis is required as the opening length in the plane of maximum membrane stress is less than the minimum length for the thickness of the shell for consideration of membrane stress calculation. Openings less than this size (diameter) are called small openings. Different sizes or equations are provided in different references. The code[1] defines small openings in vessels (not subjected to rapid fluctuations in pressure which induce fatigue) as the opening diameter (d) is less than or equal to

$$89 \text{ mm for } t \leq 10 \text{ mm or } 60 \text{ mm for } t > 10 \text{ mm}$$

Provided for two isolated openings with pitch, not less than the sum of their diameters, and in a cluster of more than two openings not less than $1+1.5\cos\alpha$ times the sum of their diameters. Where t = shell thickness and d = size of openings, α = angle, openings axis makes with cylinder long axis ($\alpha = 0$, for doubly curved shells and formed or flat heads)

Small openings may not satisfy the compensation rules for all inputs. However, this rule is valid because the hoop stress in the shell is valid for a length (L) proportional to its thickness.

8.1.3 Single Opening

Single opening is one which is neither multiple nor continuous. The compensation method for different arrangements is explained in the following sections for the cylindrical shell under internal pressure. Analysis is the same for other shells and heads except the required thickness (t_r) for pressure load. For calculating t_r, the diameter of the cone is taken as the diameter at the axis of nozzle. For partially spherical heads, the radius at the

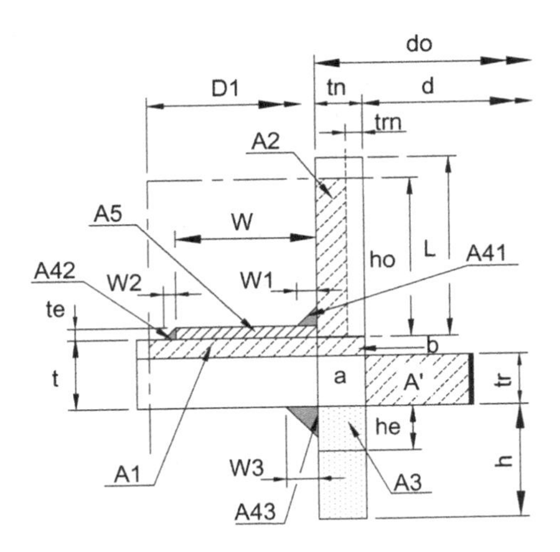

FIGURE 8.1 Cylindrical shell with an insert nozzle.

axis is considered. For ellipsoidal, the equivalent diameter (K D) is relevant. K is defined in 6.3 and equal to 0.9 for 2:1 semiellipsoidal. For flat heads, t_r is half of that required as the stress is bending for pressure and varies from zero to maximum.

The stresses (compensation) are different in planes (0° to 90°) from longitudinal (meridian) to circumferential (tangential) except the spherical shell. The maximum stressed plane depends on the type of shell and arrangement of nozzle, normally section with maximum membrane stress (longitudinal section in cylindrical shells). For cylindrical and conical shells, factor F is used for the convenience of generalizing equations as indicated in example 8.1.

For external pressure, analysis is the same as internal pressure except that the required area is half that for internal pressure because stresses are not membrane and not uniform. Furthermore, the F factor is not relevant as the thickness required to withstand pressure is the same in all planes.

8.1.3.1 Set through (Insert) Radial Nozzle

The analysis is by simple geometry for calculation of areas and explained by illustrative example 8.1. The example is for the cylindrical shell and insert radial nozzle with a pad for internal pressure as shown in Figure 8.1, followed by notes on analysis.

Example 8.1: Compensation calculations for openings in the cylindrical shell

Units: N, mm, mm^2, MPa

P = design pressure = 3

D = inside diameter of the shell = 2000

t = thickness of the shell = 28

S_V = allowed stress at design temperature of the shell = 138

d_O = outside diameter of the nozzle = 500

tn = thickness of the nozzle = 10
Sn = allowed stress at design temperature of the nozzle = 136
L = nozzle projection from the shell outside = 150
h = nozzle projection from the shell inside = 50
$W_1 = W_2 = W_3 = 10$, Weld fillet nozzle to Pad, shell to pad, shell to nozzle inside
D_P = pad diameter, effective up to D_1 only = 750
te = pad thickness = 28
S_P = allowed stress at design temperature of pad = 138
E = weld efficiency = 1
F = 1 for the longitudinal section
tr = required thickness of shell for pressure = $(P\ D/2)/(S_V\ E-0.6P) = 22$
d = diameter of circular opening = $d_O-2t_n = 480$
trn = required thickness of nozzle = $P(d_O/2-t_n)/(S_n\ E-0.6P) = 5.365$
f_1 = stress ratio of the material nozzle to shell (max 1) = $S_n/S_Y = 0.986$
f_2 = stress ratio of the material pad to shell (max 1) = $S_P/S_Y = 1$
D_1 = shell diameter limit parallel to the vessel wall = $\max(2d, d + 2t + 2t_n) = 960$
h_O = outside projection limit normal to the vessel = $\min(2.5t, 2.5t_n + t_e) = 53$
he = inside projection limit normal to the vessel = $\min(2.5t, 2.5t_n) = 25$
A_1 = area available in the shell = $\{D_1-d-2t_n(1-f_1)\}(E\ t-F\ t_r) = 2865$
A_2 = area available in the nozzle projecting outward = $2\min(h_O, L)(t_n - t_{rn})f_1 = 484$
A_3 = area available in the nozzle projecting inward = $2f_1\ t_n \min(h, h_e) = 493$
A_4 = total weld area available = $A_{41} + A_{42} + A_{43} = W_1^2 f_1 + W_2^2 f_2 + W_3^2 f_1 = 297$
A_5 = area available in the pad = $(D_P - d - 2t_n)t_e\ f_2 = 7000$
A_t = total area available = $A_1 + A_2 + A_3 + A_4 + A_5 = 11139$
A = area required for compensation = $d\ t_r\ F + 2t_n\ t_r\ F(1-f_1) = 10578$

Because $A_t > A$, design is safe.

Notes on example 8.1

1. All sizes are net after deducting negative tolerances and in corroded condition.
2. L – length shall be straight up to $2.5t_n$ from shell/pad outside.
3. F = 1 for the long section, (other sections F = 1 to 0.5, 0.5 for the circ section).
4. Weld between the shell and nozzle is full strength. Pad to shell weld strength shall be checked (refer section 8.1.7).
5. If any rigid element either side is at a distance less than its opening from centre, it will offer resistance.
6. t_r = thickness of the shell required to withstand circumferential stress due to pressure. For the longitudinal section, *tr* is half. Therefore, factor *F* (*F* = 0.5) is used in equations A_1 *and* *A* with for t_r. Because *A* is reduced by about half and A_1 increases higher than reduction in *A*, the compensation calculation in any plane other than longitudinal plane is irrelevant. Only in hill side nozzles with large offset, it is to be checked.

7. Stress ratios f_1 and f_2 are used in equations for A_2, A_3, and A_4 to reduce the actual compensating areas of nozzle and welds if the nozzle material strength is less than that of the shell ($S_Y > S_n$).
8. In case of the shell and nozzle not having the same material strength, the area ($a = 2t_r\, t_n$) of the nozzle between the inside and outside shell does not have the same strength as the shell and may require more compensating area over $d\, t_r$. Note that the actual shell material removed for inserting the nozzle is not d, but $d + 2t_n$. Furthermore, the A_1 area includes a small portion of this nozzle area [$b = (t - t_r)t_n$], and the rest of the area will resist internal pressure as the shell. Areas a and b are shown in the figure. To simplify all these effects in the calculation, the compensating area is increased by adding $2t_n\, t_r\, F(1 - f_1)$ to $d\, t_r$ in equation for A. It is zero, if the nozzle and shell materials are the same ($f_1 = 1$).
9. The shell will resist local stresses up to a limit parallel to the shell wall depending on the opening size and thickness of the shell and nozzle. The code specifies the limit each side from the nozzle outside equal to max of $d/2$ or sum of the shell and nozzle thicknesses.
10. The length limit of the nozzle outside the shell normal to shell wall which will resist local stresses is given in the code equal to 2.5 times the thickness of the shell or 2.5 times the thickness of the nozzle plus pad thickness, whichever is less outside or as well as inside of the shell. If projection is less than the above limit, only available projection is considered for compensation. Normally, the inside projection is less than the limit or no projection due to process consideration. It is to be noted that inside projection normally is not subjected to pressure and total thickness contributes in the compensating area. If corrosion is applicable due to inside fluid, it affects both sides.
11. The nozzle may be subjected to internal and external pressure on the portion inside the shell and no pressure on outside the shell like in the sleeved part of nozzle when added in case of temperature difference between the fluid in the pipe and that in the shell. Depending on the pressure condition, the value of t_{rn} may be zero or as required to external pressure. Also $t_{rn} = 0$ if the projected length is less than its thickness [refer 6.1.1a]

8.1.3.2 Set on Nozzle

The arrangement is shown in Figure 8.2. The nozzle is profile cut to suit shell OD and placed over the shell outside the surface, and the shell is opened normally with the diameter equal to nozzle ID. Opening may be provided less intentionally to reduce the area required to obtain more advantages than the increase in pressure drop.

Compensation analysis is the same as example 8.1 except A_1 and A equations are different for set on type nozzles and $W_3 = A_{43} = A_3 = 0$ (no inside projection)

A and A_1 equations are different as the factor explained in note 8 of example 8.1 is not relevant and given by

$$A_1 = (D_1 - d)(E\,t - F\,t_r)$$

$$A = d\,t_r\,F$$

For the data of example 8.1, values of areas for the set on nozzle in mm^2 are A = 10560, A_1 = 2880, and A_t = 10563. $A_t > A$; hence, design is safe.

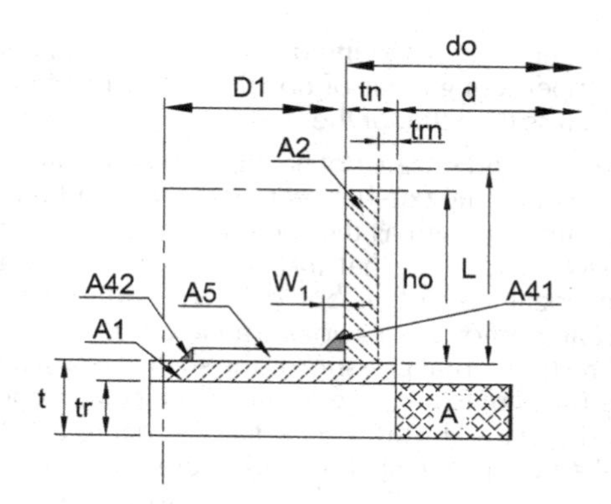

FIGURE 8.2 Cylindrical shell with the set on nozzle.

8.1.3.3 Taper Nozzle

Most of the nozzles are not entirely straight. Due to higher local stresses at the joint than those at the end, the thickness of the nozzle is provided higher at the joint and tapered down to match the mating flange or a thinner pipe usually outside to keep inside straight for smooth flow of fluid.

With the straight local effect limit being about 2.5 times, the nozzle is provided with uniform thickness up to this limit. However, for thick nozzles particularly greater than the shell thickness, the thickness is tapered down with a straight thick portion < $2.5t_n$ due to the restriction of length and/or to reduce weight. Such a nozzle as shown in Figure 8.3 is termed taper nozzle. For the purpose of analysis, the portion with a reduced thickness is considered as the nozzle and the rest of the nozzle portion hub as a pad and may not contribute fully. Only the portion up to the height t_e (equivalent thickness of pad) or h_O or 2.5t whichever is smaller is effective and acts like a pad. The value of t_e is the difference of thicknesses divided by tan30°.

The analysis is similar to example 8.1 except the variations which are given in example 8.2. The calculations in example 8.2 are with the same data of example 8.1 covering only variations.

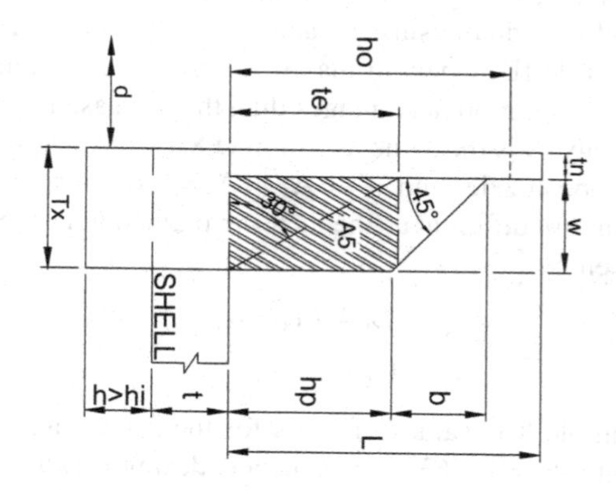

FIGURE 8.3 Taper nozzle.

8.1.3.4 Pad Type Nozzles

Hand holes are generally provided with a forged pad type flange welded over the shell with inside and outside fillet welds covered by a blind flange for closing the hand hole as shown in Figure 8.4. Compensation analysis is similar to set on type nozzle given in 8.1.3.2 with a flanged pad instead of a regular pad and no nozzle. The following example 8.3 gives variations.

Example 8.2: Remove the pad, increase the nozzle neck thickness t_X to 40mm up to a length of 45mm, and then taper down by 1:1 (45°) in arrangement of example 8.1.

New parameters in mm:

t_X = nozzle thickness at the shell juncture = 40
h_P = hub straight height = 45.

Add the following equations:

w = width of the equivalent pad = $t_X - t_n$ = 30
te = $w/\tan 30°$ = 52
h_L = hub or pad (includes bevel) area limit = min of $h_P + b$ or t_e = 52
b = bevel height = w = 30

Except the equation for the A_5 pad area for applicable equations, there is no change in other equations. The affected parameters are given below in mm, mm^2:

$h_O = \min(2.5t, 2.5t_n + t_e) = 70$
hi = $\min(2.5t, 2.5t_X)$, inside thick is t_X, not t_n = 70
$A_1 = \{D_1 - d - 2t_X(1 - f_{r1})\}(E\,t - F\,t_r) = 2860$ (t_X for t_n)
$A_2 = 597$
$A_3 = 2f_1\,t_X \min(h_i, h_e) = 3944$ (t_X for t_n)
$A_4 = W_1^2 f_1 + W_2^2 f_2 + W_3^2 f_1$, ($W_1 = 0$) = 197
A_5 (see note 3 below) = 3026
$A_t = A_1 + A_2 + A_3 + A_4 + A_5 = 10624$
$A = d\,t_r\,F + 2t_X\,t_r\,F(1 - f_1) = 10597$ (t_X for t_n)
$A_t > A$ design is safe

Notes:

1. If $h_P > h_L$, A_5 is up to limit h_L, $A_5 = 2w\,h_L$
2. If $h_L > h_P + b$, A_5 is up to $h_P + b$, A_5 = total hub area including bevel
3. Else $A_5 = 2w\,h_P$ + part of bevel $(h_L - h_P)^2$ as shown in Figure 8.3

For effectiveness of the complete hub area including bevel, reduce straight length h_P so that $t_e = h_P + b$

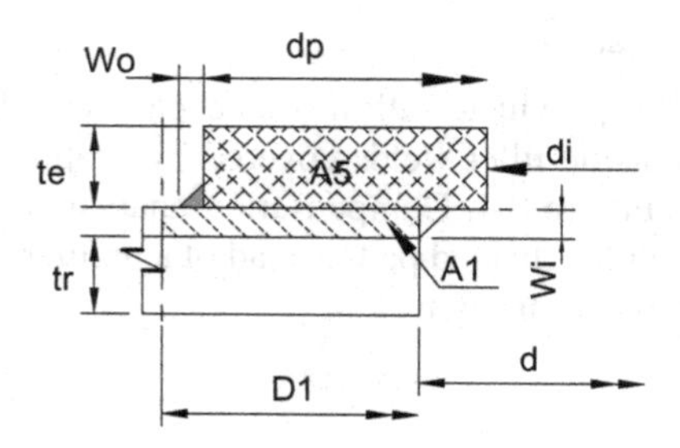

FIGURE 8.4 Cylindrical shell with a pad type nozzle.

Example 8.3: (units: N, mm) pad size 150 DN class 600, d_P = OD = 355, di = ID = 150, te = pad thick = 50, W_O = outside weld fillet = 16, and Wi = inside weld fillet = 20

$f = S_p/S_v = 136/138 = 0.986$

w = pad width = $(d_O - di)/2 = 102.5$

d = di + 2Wi = 190, d shall be as min as possible and > 150

$D_1 = 2d = 380$

$A_1 = (D_1 - d)(E\,t - F\,tr) = (389 - 190)(28 - 22) = 1140$

$A_4 = (w_o^2 + w_i^2)f_2 = 656$

$A_5 = 2w\,te\,f_2 = 2 \times 402.5 \times 50 \times 1 = 10250$

$A_t = A_1 + A_4 + A_5 = 12046$

$A = d\,tr\,F = 190 \times 22 \times 1 = 4184$

F_O = outside fillet weld strength in shear = $0.5\pi\, W_O(d_O + W_O)\, 0.49\, S_P$ = 621618

Fi = inside fillet weld strength in shear = $0.5\pi\, W_O(d + Wi)\, 0.49\, S_p = 356048$

Total fillet weld strength in shear = F_O + Fi = 977666

Required minimum strength in the welds = $S_P(A - A_1) = 414664$

Inside fillet weld required = $\min(t_n, 19) = 19 < 20$

Outside fillet weld required = $0.7 \min(t_n, 19) = 13.3 < 16$

Stud 1″ (M24), S = 172, min thread depth required = $\min[1.5, \max(1, 0.75S/S_P)]24 = 24$, and min depth of the pad undrilled shall be 6mm

8.1.3.5 Non-radial Nozzles

The nozzle attachments are not always radial. They may be tilted in any direction, mostly laterally (hill side nozzle) or axially (angular).

8.1.3.5.1 Hill Side Nozzle

It is also called offset nozzle. Move the radial nozzle by distance *B* parallel to its axis in the circumferential plane or rotate the point (C) of intersection of nozzle axis with shell mean radius (R_m) by angle α ($\sin\alpha = B/R_m$) as shown in Figure 8.5a and 8.5b, Figure 8.5a being tangential nozzle. Unlike the radial nozzle, opening diameter d is not constant and will be increasing at sections from longitudinal to circumferential. The equation for nozzle areas A_2 and A_3 is complicated as the section does not pass through the longitudinal plane of the nozzle except in the circumferential plane. At the circumferential plane, also the nozzle length within limit is higher than the radial nozzle. It may be assumed as the area within the limits h_O and h_i as shown in Figure 8.5 (as considered in ASME PTB-42013 4.5.2).

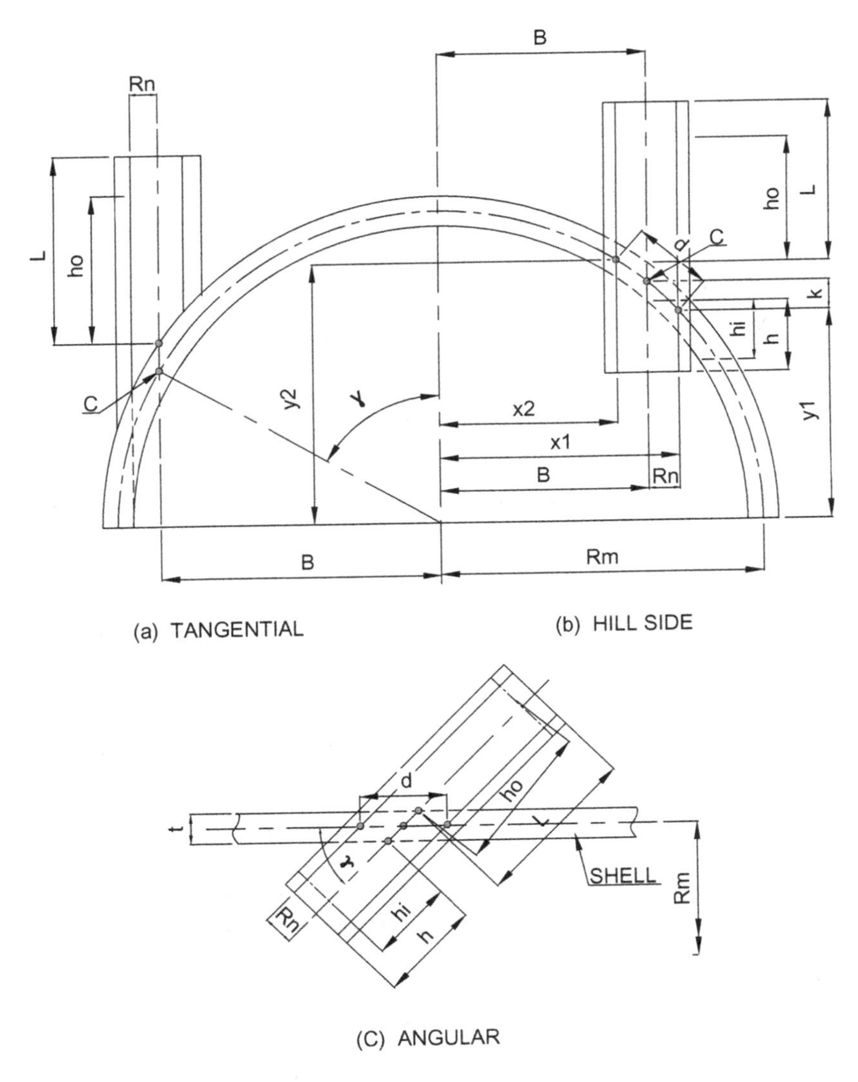

FIGURE 8.5 Non-radial nozzles.

The calculation is the same as example 8.1 except the changes explained above and given in Example 8.4 below.

Example 8.4: $\alpha = 10°$, There is no change in the *longitudinal section* (F = 1), except *L and h* will be different in both sides. Because *L and h* are much higher than limits h_O *and* h_i, there is no effect on the values of A_2 *and* A_3.

In the *circumferential section* (F = 0.5), *d* will increase, the *L and h* effect the same as above, and A_2 *and* A_3 values may change (no effect if *L and h* are higher by *k* than h_O *and* h_i). It is observed that the value of (A_t–A) for all sections other than longitudinal is higher; hence checking in other sections may not be required normally.

Increased *d* and new parameter *k* required in new equations for A_2 *and* A_3 for the circumferential section are calculated as follows (units: mm, mm²):

R_m = mean radius of the shell = (2000 + 28)/2 = 1014, effective is = $(D + t_n)/2$ = 1011

R_n = inside radius of the nozzle = 500/2 – 10 = 240

B = distance between the nozzle axis and shell axis = $R_m \sin 10°$ = 176

$x_1 = B + R_n = 416$, $x_2 = B - R_n = -64$

y = axial distance of nozzle point C from shell center = $\sqrt{R_m^2 - B^2} = 995.6$

$y_1 = \sqrt{R_m^2 - x_1^2} = 924.7$, $y_2 = \sqrt{R_m^2 - x_2^2} = 1012$

d = chord opening at mean radius = $\sqrt{(x_1 - x_2)^2 + (y_2 - y_1)^2} = 487.9$

$k = y - y_1 = 70.9$

$k' = y_2 - y = 16.4$ axial distances of nozzle point C to either side legs on Rm circle as shown in figure

Changes in D_1, A_1, and A due to the change of d and F with (E = 1, F = 0.5) are:

$D_1 = \max(2d, d + 2t + 2t_n) = 975.8$

$A_1 = \{D_1 - d - 2t_n(1 - f_1)\}(E\,t - F\,t_r) = 8287$

$A = d\,t_r\,F + 2t_n\,t_r\,F(1 - f_1) = 487.9{*}22{*}0.5 + 2{*}10{*}22{*}0.5{*}(1 - 0.9) = 5393$

$A_2 = [\min(L + k, h_O) + \min(L - k, h_O)](t_n - t_{rn})f_1 = 484$

$A_3 = f_1\,tn[\min(h - k, hi) + \min(h + k', hi)] = 0.986 \times 10 \times (0 + 25) = 246.5$

Note that there is no change in A_2 because $h_o < h - k$, but A_3 decreases as $h - k < h_i$

8.1.3.5.2 *Angular Nozzle*

Rotate the radial nozzle axis in the long plane of the shell at its intersection with the mean radius of the shell by angle as shown in Figure 8.5c.

There is no change in calculation for the *circ section* (F = 0.5). In the *longitudinal section*, *d* will increase by 1/cosα. Other considerations are the same as hill side and radial nozzles.

8.1.4 Multiple Openings

Whenever the distance between two openings (other than continuous openings) is less than half of the sum of the limits of each in the shell (D_1 in example 8.1 above) as shown in Figure 8.6, the opening pair is called multiple opening. In such case, the

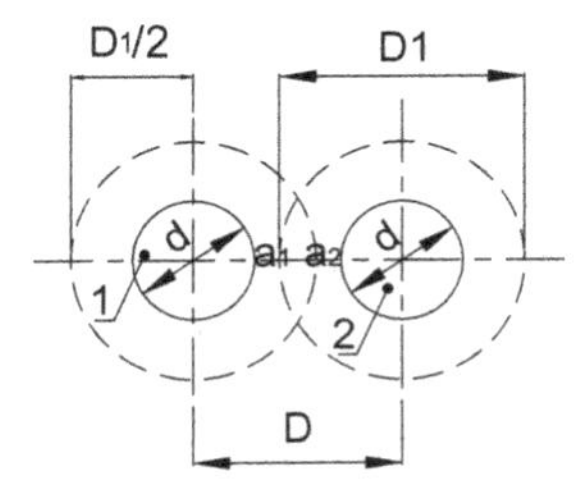

FIGURE 8.6 Multiple openings.

limits are to be reduced in proportion to their openings. Pitch less than 4/3 of average diameter of multiple openings (d) is considered as continuous openings and may lead to ligament failure. For circumferential multiple openings, the limiting angle between openings can be calculated as D_1/R_m radians and R_m = shell mean radius.

Local stresses due to openings under internal pressure are maximum at the shell–nozzle longitudinal juncture points either side of opening and gradually reduce away and can be considered as insignificant after certain limit ($D_1/2$) from its center. In multiple openings (*1 and 2*), the distance between adjacent openings (D) is less than the limit D_1, due to which the local stresses at a_1 from opening-2 and at a_2 from opening-1 are not insignificant. Therefore, stresses are at max point (junction) of each opening (a_1 or a_2), stresses are added from other side openings, and ligament failure may result even though compensation can be satisfied by providing the major area from nozzles. To strengthen the ligament portion, the code prescribes at least 50% of area required shall be provided by the shell only. Furthermore, it is also possible to comply this with much lesser ligament, but if stress at a_1 *and* a_2 has already reached the allowed limit, the additional stresses from other sides are significant; thereby stresses at these two points may exceed the allowed limit. Therefore, minimum ligament is required so that the addition of stresses at each juncture point from other sides is insignificant. The code prescribes a minimum pitch of 1.33 times the average of both openings.

8.1.5 Continuous Openings (Ligaments)

Continuous openings (more than two unreinforced or two with pitch less than 1.33 times of their average diameter) in long or circ or any direction as shown in Figure 8.7 are analyzed by ligament method illustrated below. The stresses may exceed allowable even if the size of openings is small as defined in 8.1.2. Example 8.5 will explain the above and define and calculate the ligament efficiency.

p = longitudinal pitch

d = diameter of opening in the shell

Net resisting length per one pitch = p–d

Pressure width per one pitch = p

D = mean diameter of the shell

t = shell thickness

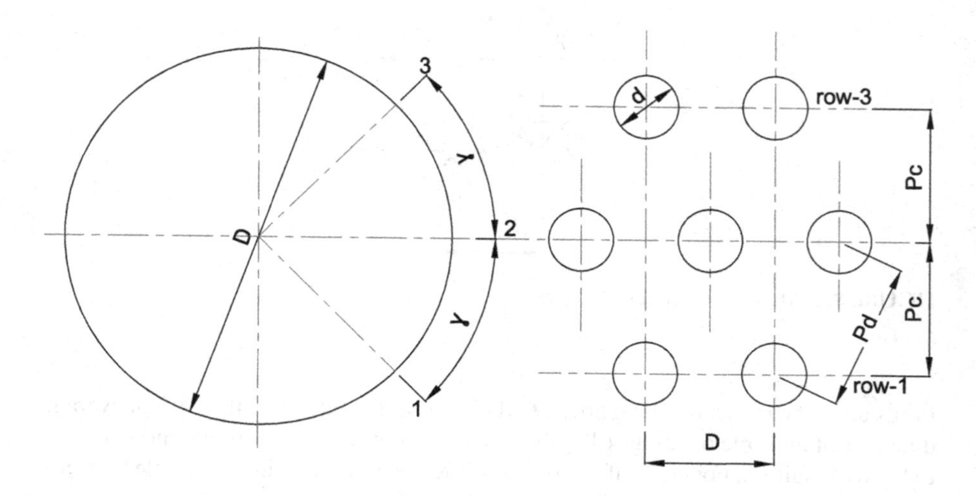

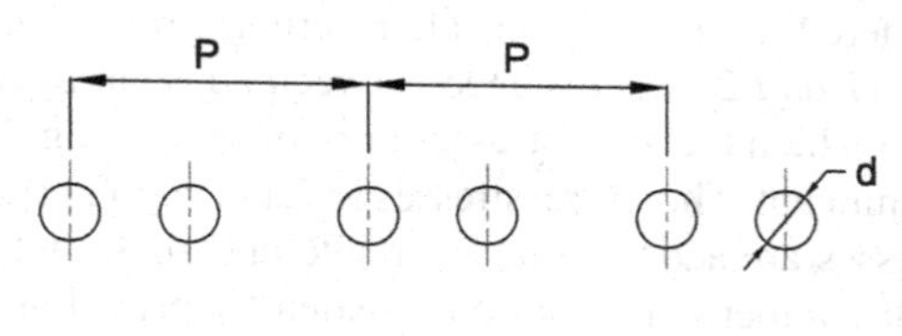

FIGURE 8.7 Continuous openings.

Using membrane theory force in the circumferential direction

$$P\,D\,p = S\,2t(p - d)$$

Rewriting, $t = P\,D/[2S(p-d)/p] = P\,D/(2S\,E)$

Example 8.5: Calculate ligament efficiency for various tube layouts (Figure 8.7):

1. single row: $E = (p - d)/p$
2. single row unequal pitch: $E = (p - n\,d)/p$ (Figure 8.6b)
3. more than one row: *E* can be computed as explained in example below.
   D = mean diameter of shell = 1000 mm
   p = long pitch = 100 mm
   p_C = circumferential pitch = $\pi\,D(\alpha/360) = 70$ ($\alpha = 8°$)
   d = diameter of opening (ID of nozzle) = 50 mm

p_d = diagonal pitch = $\sqrt{(p^2/4 + p_C^2)}$ = 86 mm
Long $E_L = (p - d)/p = 0.5$ (used in hoop stress)
Circ $E_C = (2p_C - d)/2p_C = 0.666$ (used in long stress)

Because long stress is half of circ stress, when $E_C = E_L/2$, both long and circ stresses will be equal. Therefore, equivalent long efficiency E_{ce} is given by

$E_{ce} = 2\ E_C = 2*0.666 = 1.332$
Diagonal $E_d = (p_d - d)/p_d = 0.42$
E_{de} = Equivalent diagonal efficiency for E_d is, nearer to $2E_d$ when p_d is nearer to p_c, and nearer to E_d when p_d is nearer to $p/2$. That is, E_{de} varies from E_d to $2E_d$ and is given by empirical formula (Ref. 1)
$E_{de} = 0.01\,[J + 0.25-(1-E.long)\sqrt{(0.75 + J)}]/(0.00375 + 0.005J)) = 0.5$

Where $J = (p_d/p)^2 = 0.74$

E minimum equivalent ligament efficiency = min(E_l, E_{ce}, E_{de}) = 0.5
Considering pressure P = 1 MPa, all. Stress = 100 MPa, t = 8 mm
Required thickness for unpierced shell = P D/2S = 1*1000/200 = 5 mm and safe.
Required thickness in the above example = P D/(2S E) = 1*1000/(2*100*0.5) = 10 mm and not safe.

As per 8.1.2, it is small opening and required t = 5 mm and safe.
The conclusion is that the small opening is applicable only for a single isolated opening but not to continuous openings.

8.1.6 Large Bore

The size of large opening (ID of nozzle d) is limited depending on inside diameter (D) of the shell, beyond which local stresses higher than allowable may set up even after the compensation rule is satisfied. Because the required compensation area can be provided by the increasing nozzle thickness local to junction with the thickness of the shell just sufficient to resist pressure, the discontinuity stresses and stress concentration in the shell may be large enough to exceed allowed. Therefore, for the d/D ratio beyond a limit, it is required to provide part of compensation required in the shell within closer limits. Furthermore, it is safer to calculate local membrane and membrane plus bending stresses. Appendix 1.7 of code[1] gives two different limits of compensation and equations for membrane and bending stresses in both limits and is explained below with examples. Some codes give rules for margin in shell thickness for local stresses in all single openings.

The equation for d/D is given as

$$d = \text{If } D > 1500 \text{ mm, Min}(D/3, 1000), \text{ else Min}(D/2, 500)$$

The following analyses 1 and 2 are to be satisfied for shell and large bore nozzle arrangement to be safe under internal pressure. If [D > 1020, and d > 1020 and $3.4\sqrt{(0.5D\ t)}$, and $d/D \leq 0.7$], analysis 3 is to be satisfied.

1. Compensation method used in example 8.1
2. A min compensation area is required in both the shell and nozzle to resist discontinuity stress. $0.75A_1$ and A_2 shall be at least equal to $2/3A$, where A, A_1, *and* A_2 are as calculated in example 8.1.
3. $P_m \leq S$ and $P_m + P_b \leq 1.5S$ computed with illustrative example 8.6 below.

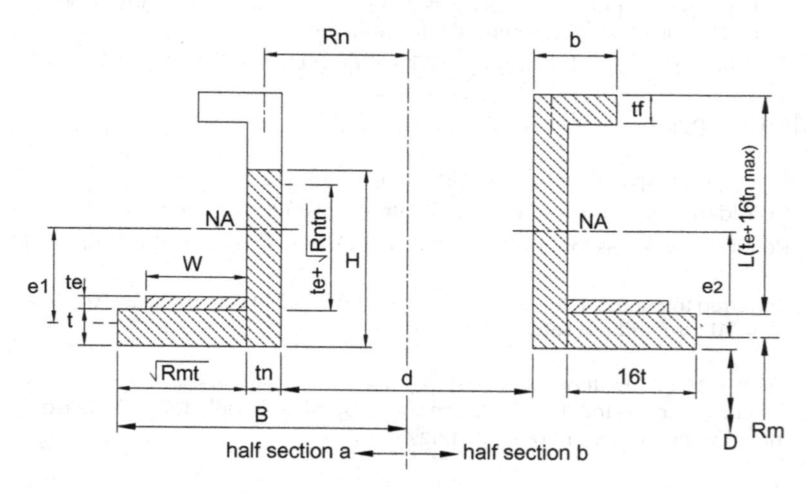

FIGURE 8.8 Large openings.

Example 8.6: Notation as in example 8.1 and Figure 8.8, units: N, mm, MPa

The shell is the same as in example 8.1, pressure is less P = 1 MPa, and nozzle is 1080 OD and 28 thick and 200 long and has 125 × 28 pad

$D = 2000$, $d_O = 1080$, $w = 125$, $t = tn = te = 28$, $L = 200$, $b = 100$, $R_m = (2000 + 28)/2 = 1014$, $Rn = (1080–28)/2 = 526$, $d = d_O - 2tn = 1024$, $S = 138$

For Figure 8.8a:

P_m = membrane stress = $P\,A_p/A_r$

P_b = bending stress M/Z

Ar and A_P are half of resisting and pressure areas as only the half section is considered for convenience; the boundaries for areas are

Normal to shell from shell OD: $t_e + \sqrt{R_n t_n} = 28 + 121.4 = 149.4$

Parallel to shell: $\sqrt{R_m t}$ from nozzle OD = 168.5

$B = 0.5d_O + \sqrt{R_m t} = 0.5*1080 + 168.5 = 708.5$

$H = t + t_e + \sqrt{R_n t_n} = 28 + 28 + 121.4 = 177.4$

$A_P = D\,B + d\,H = 2000*708.5 + 1024*177.4 = 1598658$

$Ar = 2t(B - d_O/2) + 2t_n\,H + 2t_e \min[w, \sqrt{R_m t}]$

$= 2*28(708.5-1080/2) + 2*28*177.4 + 2*28*125 = 26370.4$

$P_m = PA_p/A_r = 1*1598658/26370.4 = 60.6 < S$

$Za = 218916.7$, $e = 47.58$, Z of area A_r and NA from shell mean as shown in Figure 8.8a

$Ma = P(d^3/48 + 0.25D\,de_1)]$ (refcode) $= 1(1024^3/48 + 0.25*2000*1024*47.58)$ $= 46730581$

P_{b1} = bending stress $M_a/Z_a = 46730581/218916.7 = 213$

For Figure 8.8b: flange dimensions are $b \times t_f = 100 \times 28$ and boundaries are:

Normal to shell: $t_e + \min(L, 16t_n)$ from shell OD $= 28 + 16*28 = 476$, limited to $L = 200$

Parallel to shell: $16t$ from nozzle OD $= 16*28 = 448$

$Z_b = 866181$, $e_2 = 65.4$, and Z of area A_r and NA from the shell mean as shown in Figure 8.8b

$M_b = P(d^3/48 + 0.25D\,de_2)]$ (refcode) $= 1(1024\text{^}3/48 + 0.25*2000*1024*65.4)$ $= 23780962$

P_{bb} = bending stress $M_b/Z_b = 23780962/866181 = 27.5$

$P_m + P_b = P_m + \max(P_{ba} + P_{bb}) = 60.6 + 213 = 273.6$

Stress exceeds allowable 1.5S, and by increasing the pad width to its limit 168.5, the stresses can be reduced. If it still exceeds, the thickness of the pad and/or nozzle can be increased. Increasing the shell thickness is more effective but not economic as increase cannot be limited to required length.

Reinforced (compensated) openings in formed head can be any size, but advisable to be less than half of the diameter.

8.1.7 Weld Strength

The shell, nozzle, and pad (if provided) shall be integral so that the compensation is effective. Shell to nozzle and pad to nozzle can be considered as integral with defect-free full weld, but pad to shell weld cannot be full welded and only fillet weld is practical. A fundamental requirement is that there shall be no relative displacement between these three parts. Hence, partial welds can be used provided welds are sufficient to give required strength. Therefore, pad to shell weld and nozzle to shell and nozzle to pad if not full welded needs to be checked. Analysis is by structural basics except the weld factor is taken from the code.

There are three paths of failure as shown in Figure 8.9 for set through nozzle joint with the pad dealt in example 8.1. The elements of failure path are

a. pad to shell fillet weld (A_{42})
b. pad to nozzle fillet weld outside (A_{41})
c. pad to nozzle groove weld outside
d. shell to nozzle groove weld in the opening
e. shell to nozzle fillet weld inside (A_{43})
f. nozzle wall

The failure paths are

1. shell axial path: a + f
2. nozzle axial path: b + c + d + e
3. pad to shell: a + d + e

If the pad is not provided, the third path is not applicable, and in other two paths, pad weld is not applicable. The same logic is taken for any type of weld joints.

Load is the same as in the calculation of hoop stress. The element *a* through *f* will transfer the load to shell. In other words, all the compensating areas except that of the shell multiplied by their allowable stress are loads. Strength of elements shall be adequate to resist the loads and due to the cross section area each side of compensation plane.

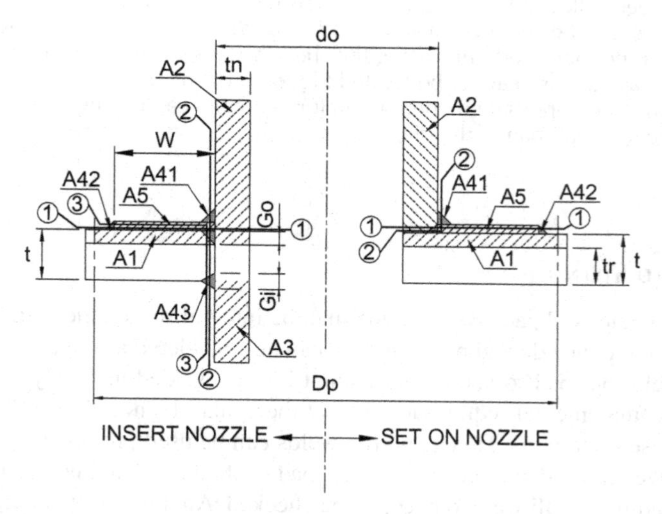

FIGURE 8.9 Weld strength in the shell nozzle joint.

The effects of pressure and resistance of welds for the insert nozzle are illustrated in example 8.7.

Example 8.7: The data are the same as the example in Table 8.1 except the full weld between the shell and nozzle replaced with groove welds outside (G_O = 11 mm) and inside (G_i = 9 mm) and pad to nozzle (Gp = 0).

Notation as per example 8.1 and Figure 8.1

Weld loads on element welds or nozzle wall in N:

Load carried by any element in the path = (compensating areas transferring load to that element) × S_V

W_{11} = load on Weld in Path1 = $(A_2 + A_5 + A_{41} + A_{42})S_V$ = (484 + 7000 + 98.6 + 100)138 = 1060198 N

$2t_n\, t\, f_1$ = 2*10*28*0.986 = 552

W_{22} = load on Weld in Path2 = $(A_2 + A_3 + A_{41} + A_{43} + 2t_n\, t\, f_1)\, S_V$ = (484 + 493 + 98.6 + 100 + 552)138 = 238430, $(A_2 + A_{41})S_V$ for set on

W_{33} = load on Weld in Path3 = $(A_2 + A_3 + A_5 + A_4 + 2t_n\, t\, f_{r1})\, S_V$ = (484 + 493 + 7000 + 297 + 552)138 = 1220770, $(A_2 + A_4)S_V$ for set on

W = total load on Welds = $[A - A_1 + 2t_n\, f_{r1}(E_1 t - Ft_r)]S_V$ = [10578 – 2865 + 2*10*0.986(1*28 – 1*22)]138 = 1080651, $(A - A_1)S_V$ for set on

Strength of elements of failure path analysis in N:

Strength (resistance) of element in N is the product of the following three parameters.

1. Area of element each side = half of perimeter × effective thickness of weld or nozzle (0.7W for fillet, G for groove, and t_n for nozzle)
2. Allowed stress of element minimum of connecting elements
3. Weld factor = 0.74 for tension and = 0.7 for shear (ref code)

The following are element resistances in N:

S_a shear in shell to nozzle outside fillet (A_{41}) = $[(\pi/2)d_O\, 0.7W_1]\, S_n\, 0.7$ = 523391

S_b shear in shell to pad fillet (A_{42}) = $[(\pi/2)D_P\, 0.7W_2]$ Sp 0.7 = 796631

S_C shear in shell to nozzle inside fillet (A_{43}) = $[(\pi/2)d_O\, 0.7W_3]\, S_n\, 0.7$ = 523391

S_d shear in the nozzle wall = $[(\pi/2)(d_O - t_n)tn]\, S_n\, 0.7$ = 732747

T_a tension in the shell outside groove = $[(\pi/2)d_O\, G_O]\, S_n\, 0.74$ = 962467

T_b tension in pad groove = $[(\pi/2)d_O\, G_P]S_P\, 0.74$ = 0

T_C tension in the shell inside groove = $[(\pi/2)d_O\, G_i]S_n\, 0.74$ = 721845

Total strength of combination of elements in each path:

S_1 = Path11 = $S_b + S_d$ = 1529378 N

S_2 = Path22 = $S_a + T_e + T_f + T_g + S_C)$ = 3613344 N
S_3 = Path33 = $S_b + T_f + T_g + S_C)$ = 2650883 N

Summary of failure Path Calculations

S_1 > W1-1 or W, 1529378 > 1060198 or 1080651
S_1 > W2-2 or W, 2650883 >213784 or 1080651
S_1 > W3-3 or W, 2924123 >1220770 or 1080681

8.1.8 Local Stresses due to Opening in Other Shells and Heads

The basics are the same as explained for the cylindrical shell above. The common change in calculations in sub-sections of 8.1 is the equation for t_r, which is as per type of shell. For formed heads with varying radius, the average radius of cut out opening is relevant. For 2:1 semiellipsoidal, 0.9D is used.

Changes with respect to the type of shell are:

1. *Cone*, the same as cylinder converting its parameters into the equivalent cylinder as explained in 7.6.1.1.
2. *Spherical including elliptical or partially spherical* F is not applicable as all planes are identical. Remove *F* or $F = 1$. Limit large bore diameter to $D/2$.
3. *Flat closure*: *F* is not applicable as all planes are identical. Remove *F* or $F = 1$. t_r is due to bending stress and varies (min 0) depending on the location of opening, t_r in equation A & A_1 is taken as $t_r/2$ (average). Limit opening to $D/2$. Compensation calculation given above is not applicable when the flat plate is strengthened by stiffeners or by staying for pressure load.

8.1.9 Opening in Shells Subjected to External Pressure

The compensation method is the same as explained for internal pressure except the following.

1. The reinforcement area required for openings in shells subjected to external pressure needs only 50% that of internal pressure as the stress is bending (not membrane).
2. Min required thickness of the shell and nozzle is under external pressure.
3. The value of *F* shall be 1.0 as stress is the same in any plane.
4. There is no change for the flat head with respect to the required area.

8.2 LOCAL STRESSES AT THE JUNCTION IN THE SHELL DUE TO EXTERNAL FORCES THROUGH ATTACHMENTS INCLUDING NOZZLES

External force tensor consists of three forces P, V_C, and V_L and three moments *M* with suffix t, L, and c. Suffixes of forces and moments are based on their effect.

where

c = circumferential (circ), L = longitudinal (long), and t = torsion

Force tensor in the order of right hand thumb rule = $\{V_C, V_L, P; M_L, M_C, M_t\}$ similar to $\{F_X, F_Y, F_Z; M_X, M_Y, M_Z\}$, convention in this section is as used in WRC bulletins. P is + if force is towards the shell and – outwards. Stresses are tensile + and – compression.

On the perimeter of joint, max combined stress due to force tensor and pressure can be at longitudinal or circumferential section, and can be at inside or outside of thickness. Stresses at other points on perimeter will be less than maximum. Therefore, stress is to be calculated at eight points in each shell and nozzle as shown in Figure 8.10.

Eight points are denoted by A_o, A_i, B_o, and B_i (long); C_o, C_i, D_o, and D_i (circ)

Where suffix i = inside and o = outside.

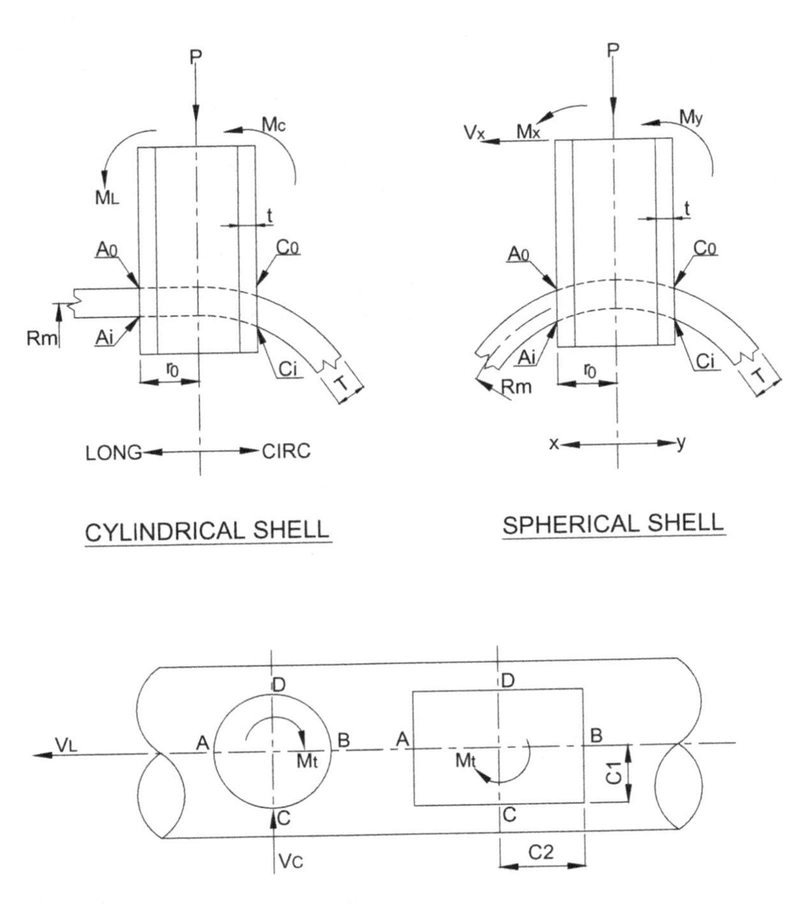

FIGURE 8.10 Local stresses in the shell.

At each point shear, membrane and bending (±) stresses in both circ and long planes are induced due to external force tensor apart from general membrane stress due to pressure.

P induces the same negative membrane (m) at all eight points and bending stress (b) inside + & outside – at A and B, and similar at C (as A) and D (as B) in each of long and circ directions.

V_C induces the same shear stress (τ_1) at A and B but + at A and – at B

V_L induces the same shear stress (τ_2) at C and D but + at D and – at C

M_t induces the same torsional shear stress (τ_3) – sign at all eight points

M_C induces the same membrane (m) – at D & + at C and bending (b) stresses + at C_O and D_i and – at Ci and D_O.

M_L induces the same membrane (m) – at A and + at B and bending (b) stresses + at B_O and Ai and – at Bi and A_O.

Except P and M_t (axial force and moment), the direction of other four will not influence combined stress. Normally, the stress due to M_t is insignificant compared to others, and the direction of force *P* has large influence in combined stress. Because pressure stress is tensile and positive at all eight points, combined stress will be higher if stress due to *P* is also positive (stress is + for –P).

Shear stress τ and pressure stress σ_C and σ_L are calculated using simple equations explained in previous chapters. Other stresses are computed using the following steps

Step 1: Computation of local stress membrane (m) and bending (b)

m and *b* can be expressed by basic equations $m = N/T$ and $b = M/Z = 6M/T^2$

These stresses are expressed by Eqs. 8.1 and 8.2 for the purpose of using graphs in WRC bulletins.

$$m = N/T = (N/C_n)(C_n/T) \tag{8.1}$$

$$b = 6M/T^2 = (M/C_b)(6C_b/T^2) \tag{8.2}$$

$$\text{Membrane} + \text{bending stress } m + b = K_n N/T \pm K_b 6M/T^2$$

where

T = thickness of the shell

K_n and K_b are stress intensification factors, normally $K_n > K_b$, and both are neglected in this book. K_n and K_b values depend on the ratio of thickness to fillet radius. With a large fillet radius, K_n and K_b can be limited to 1.

N and M are force and moment per unit circumference.

N/C_n and M/C_b are dimensionless coefficients for P, M_c, and M_L and plotted against shell and attachment parameters from Bijlaard's data (WRC 107) and Steele's data (WRC 297) and can be taken from WRC107(537) & 297 bulletins.

C_n and C_b are functions of P or M_c or M_L and geometrical parameters used for the purpose of graphs. C_n and C_b are computed for various arrangements in respective sections.

Suffix ø for circ stress and *r* for long stress are used with N and M in the above stress equations.

***Step* 2**: Compilation of combined membrane ($P_m + P_L$) and membrane + bending ($P_m + P_L + Q$) stresses.

All the stresses compiled in step 1 are added, ($P_m + P_L$) and ($P_m + P_L + Q$) are calculated at all eight points, and max of each is limited to allowed. Allowed stresses are1.5S and 3S as the stresses are local and secondary stress category (refer section 4.4.1.2)

C_m = sum of all circ membrane stresses + hoop stress

L_m = sum of all long membrane stresses + long stress

C = sum of all circ membrane + bending stresses + hoop stress

L = sum of all long membrane + bending stresses + long stress

τ_m = shear stress due to V_C and V_L

τ = shear stress due to M_t, V_C, and V_L

Using basics (section 3.7.3.4) of combining stresses C, L, τ combined stress is given by

$$0.5\{\mathrm{abs}(C+L)+\sqrt{(C-L)^2+4\tau^2}\}$$

Considering the effect of sign of each stress and the following equations are compiled to derive max stress with all combinations.

$$m_1 = \sqrt{(C_m - L_m)^2 + 4\tau_m^2} \tag{8.3}$$

$$P_1 = \mathrm{if}[C_m/L_m < 0, m_1, \mathrm{else}\ 0.5\{\mathrm{abs}(C_m + L_m) + m_1\}] \tag{8.4}$$

$$p_m + p_L = \mathrm{if}\left[\tau_m = 0, \max\{C_m, L_m, \mathrm{abs}(C_m - L_m)\}, \mathrm{else}\ P_1\right] \le 1.5S \tag{8.5}$$

$$b_1 = \sqrt{(C-L)^2 + 4\tau^2} \tag{8.6}$$

$$P_2 = \mathrm{if}\ [C/L < 0, b_1, \mathrm{else}\ 0.5\ \{\mathrm{abs}(C+L) + b_1\}] \tag{8.7}$$

$$p_m + p_L + Q = \mathrm{if}\left[\tau = 0, \max\{C, L, \mathrm{abs}(C-L)\}, \mathrm{else}\ P_2\right] \le 3S \tag{8.8}$$

C/L < 0 means that the sign of C and L is not the same. It may be observed that the shell may withstand loads up to a certain limit irrespective of meeting its hoop stress and compensation requirement due to pressure. Therefore, resisting combined load of pressure and external loads by the above local load analysis does not guarantee that the shell is safe for pressure. It shall have sufficient thickness to withstand hoop stress as well as additional primary membrane and bending stresses due to opening under pressure.

8.2.1 Cylindrical Shell

Difference with WRC107: suffixes: o, i, and r are used in place of u, L, and x of WRC

8.2.1.1 Cylindrical Attachment or Nozzle

Local stresses are calculated using the procedure explained in 8.2. Equations for Cn and C_b are given below

$$Cn = P/R_m,\ M_C/(R_m^2\ \beta) \text{ and } M_L/(R_m^2\ \beta)$$
$$C_b = P,\ M_C/(R_m\beta) \text{ and } M_L/(R_m\beta)$$

The dimensionless parameters ($N_ø$ or N_r)/Cn, ($M_ø$ or M_r)/C_b in both long and circ planes can be read from graphs or tables against shell parameter $\gamma = R_m$ T and attachment parameter $\beta = 0.875r_O/R_m$ from graphs in WRC107/537.

The detailed calculations are illustrated by examples in Table 8.1. If the pad is provided, it is taken as an attachment and analyzed separately.

The attachment or nozzle is assumed as rigid compared to the shell. Hence, there are no local stresses in the attachment. When nozzle is not rigid, stresses in shell increase and nozzle stresses are considerable and covered in 8.3.

8.2.1.2 Square/Rectangular Attachment

Analysis is the same as that for the cylindrical attachment except the attachment parameter β which is evaluated by

Square attachment $\beta = c/R_m$, where c = width of square

Rectangular attachment with half width C_1 = in circ direction and C_2 = in long direction. β varies and is given below for each load:

a. For radial load P:

$$\text{If } C_1 \geq C_2,\ \beta = [1-(C_1/C_2-1)(1-K_1)/3]\sqrt{(C_1C_2/Rm^2)}$$
$$\text{If } C_1 < C_2,\ \beta = [1-4(1-C_1/C_2)(1-K_2)/3]\sqrt{(C_1C_2/R_m^2)}$$

K_1 and K_2 are given in Table 6 of WRC 107

b. For circumferential moment M_C:
   1. For membrane stress $\beta = \{(C_1^2\ C_2)\}^{1/3}/R_m$, and the value obtained from the relevant graph is to be multiplied by correction factor C_C (C_C from Table 7 of WRC 107)
   2. For bending stress $\beta = K_C\{(C_1^2C_2)\}^{1/3}/R_m$, and K_C is given in Table 7 of WRC 107

TABLE 8.1
Calculation of Local Stresses in the Cylindrical Shell Due to External Force Tensor on the Attachment (cyl. nozzle) Under Internal Pressure
Shell material SA516 70, units: N, mm, MPa

1) Input data

Rm = mean radius of cylindrical shell	1014
T = thickness of shell	28
r_O = outside radius of attachment	250
p = internal pressure	3
S = allowable stress at temp 250°C	138
Kn & K_b stress intensification factors not considered	

2) External loads

P = radial(axial) load =	100000
V_L = longitudinal load =	100000
V_C = circumferential load =	100000
Mt = torsional moment =	1E + 07
M_C = circ. moment =	1E + 07
M_L = longitudinal moment	1E + 07

3) Stresses due to pressure

σc-circ stress = p R_m/T = 108.6

σ_L-longl stress = σ_C/2 = 54.3

4) Cn, C_b to calculate stresses due to Fa, M_C, M_L

C_{an}	C_{ba}	C_{nC}	C_{bC}	C_{nL}	C_{bL}
P/Rm	P	$\frac{M_C}{R_m^2\beta}$	$\frac{Mc}{R_m\beta}$	$\frac{M_L}{R_m^2\beta}$	$\frac{M_L}{R_m\beta}$
98.619	100000	45.03	45657	45.03	45657

5) Geometrical shell & nozzle parameters for input to graphs

γ = Rm/T = 1014/28	Min 5, max 300	36.21
β = 0.875r_O/Rm = 0.875 × 250/1014	Min 0.01, max 0.5	0.216

6) Output dimensionless constants from graphs [Ref. WRC-537(107)][2]

Circumferential stress in long Plane

	Fig	Value
Nø/Cn = Nø/(P/Rm)	4C	4.937
Mø/C_b = Mø/P), (A,B)	2C1	0.032
Nø/Cn = Nø/[M_L/(Rm²β)]	3B	3.570
Mø/C_b = Mø/[M_L/(Rmβ)]	1B1	0.024

Longitudinal stress in long Plane

	Fig	Value
Nr/Cn = Nr/(P/Rm)	3C	3.076
Mr/C_b = Mr/P (at A, B)	1C1	0.064
Nr/Cn = Nr/[M_L/(Rm² β)]	4B	1.483
Mr/C_b = Mr/[M_L/(Rm β)]	2B1	0.040

Circ. stress in circ. Plane

	Fig	Value
Nø/Cn = Nø/(P/Rm),(at C,D)	3C	3.076
Mø/C_b = MøP,[C,D(max 0.5)]	1C	0.063
Nø/Cn = Nø/[Mc/(Rm²β)]	3A	1.486
Mø/C_b = Mø/[Mc/(Rmβ)]	1A	0.076

For long. stress in circ. Plane

	Fig	Value
Nr/Cn = Nr/(P/Rm),(at C, D)	4C	4.937
Mr/C_b = Mr/P [at C, D(max 0.3)]	2C	0.033
Nr/Cn = Nr/[M_C/(Rm²β)]	4A	2.894
Mr/C_b = Mr/[M_C/(Rmβ)]	2A	0.036

(Continued)

TABLE 8.1 (*Continued*)

Calculation of Local Stresses in the Cylindrical Shell Due to External Force Tensor on the Attachment (cyl. nozzle) Under Internal Pressure

7) Stress tables due to P, Mc, and M_L

m-membrane stress = Eq. 8.1, b-bending stress = Eq. 8.2, Cm-total circumferential membrane stress, C-total circ. bending stress, Lm-total longitudinal membrane stress, L-total longitudinal bending stress

Circumferential stress in the longitudinal Plane					**Circ. stress in circ. Plane**			
At points	**Ao**	**Ai**	**Bo**	**Bi**	**Co**	**Ci**	**Do**	**Di**
1. m(P) = eq. 8.1	−17.39	−17.39	−17.39	−17.39	−10.83	−10.83	−10.83	−10.83
2. b(P) = eq. 8.2	−24.49	24.49	−24.49	24.49	−48.21	48.21	−48.21	48.21
3. m(Mc) = eq. 8.1	0.00	0.00	0.00	0.00	−2.39	−2.39	2.39	2.39
4. b(Mc) = eq. 8.2	0.00	0.00	0.00	0.00	−26.56	26.56	26.56	−26.56
5. m(M_L) = eq. 8.1	−5.74	−5.74	5.74	5.74	0.00	0.00	0.00	0.00
6. b(M_L) = eq. 8.2	−8.39	8.39	8.39	−8.39	0.00	0.00	0.00	0.00
Cm = 1 + 3 + 5 + σc	85.51	85.51	97.00	97.00	95.42	95.42	100.20	100.20
C = Σ(1to 6) + σc	52.64	118.39	80.89	113.10	20.65	170.19	78.54	121.86
Longitudinal stress in long. Plane					**Long. stress in circ. Plane**			
1. m(P) = eq. 8.1	−10.83	−10.83	−10.83	−10.83	−17.39	−17.39	−17.39	−17.39
2. b(P) = eq. 8.2	−48.98	48.98	−48.98	48.98	−25.26	25.26	−25.26	25.26
3. m(Mc) = eq. 8.1	0.00	0.00	0.00	0.00	−4.65	−4.65	4.65	4.65
4. b(Mc) = eq. 8.2	0.00	0.00	0.00	0.00	−12.58	12.58	12.58	−12.58
5. m(M_L) = eq. 8.1	−2.38	−2.38	2.38	2.38	0.00	0.00	0.00	0.00
6. b(M_L) = eq. 8.2	−13.98	13.98	13.98	−13.98	0.00	0.00	0.00	0.00
Lm = 1 + 3 + 5 + σL	41.10	41.10	45.87	45.87	32.28	32.28	41.59	41.59
L = Σ(1 to 6) + σL	−21.85	104.06	10.87	80.88	−5.56	70.11	28.91	54.26
8) Shear stress due to Vc, V_L, and Mt								
τ – total shear, τ_m – shear membrane, $\tau_1 = Vc/(\pi r_O T)$, $\tau_2 = V_L/(\pi r_O T)$, $\tau_3 = Mt/(2\pi r_O^2 T)$								
Normal to longitudinal plane for τ					**Normal to circ plane for τ**			
τ_1 due toVc	4.55	4.55	−4.55	−4.55	0.00	0.00	0.00	0.00
τ_2 due to V_L	0.00	0.00	0.00	0.00	−4.55	−4.55	4.55	4.55
τ_3 Mt	0.91	0.91	0.91	0.91	0.91	0.91	0.91	0.91
$\tau = \tau_1 + \tau_2 + \tau_3$	5.46	5.46	−3.64	−3.64	−3.64	−3.64	5.46	5.46
$\tau m = \tau_1 + \tau_2$	4.55	4.55	−4.55	−4.55	−4.55	−4.55	4.55	4.55
9) Combined membrane stress = Pm + P_L = max(eight points)								
Cm – Lm	44.41	44.41	51.12	51.12	63.14	63.14	58.61	58.61
Cm + Lm	126.62	126.62	142.87	142.87	127.70	127.70	141.79	141.79
m_l = eq. 8.3	45.33	45.33	51.93	51.93	63.79	63.79	59.31	59.31
P_l = eq. 8.4	85.97	85.97	97.40	97.40	95.74	95.74	100.55	100.55
Pm + pl = eq. 8.5	86.0	86.0	97.4	97.4	95.7	95.7	100.5	100.5
10) Combined membrane + bending stress = Pm + P_L + Q = max(eight points)								
C – L	74.49	14.33	70.02	32.22	26.20	100.08	49.63	67.59
C + L	30.78	222.45	91.76	193.97	15.09	240.30	107.45	176.12
b_l = eq. 8.6	75.29	18.01	70.40	33.04	27.20	100.34	50.81	68.47
P_2 = eq. 8.7	75.29	120.23	81.08	113.50	27.20	170.32	79.13	122.29
Pm + p_L + Q = eq. 8.8	75.29	120.23	81.08	113.50	27.20	170.32	79.13	122.29
Max stresses	pm + p_L	100.549	<all = 207	pm + p_L + Q	170.32	<all = 414	Result	Safe

c. For longitudinal moment M_L:
   1. For membrane stress $\beta = \{(C_1C_2^2)\}^{1/3}/R_m$ and the value obtained from the relevant graph is to be multiplied by correction factor C_L (C_L from Table 8 of WRC 107)
   2. For bending stress $\beta = K_L\{(C_1\ C_2^2)\}^{1/3}/R_m$ and K_L is given in Table 8 of WRC 107

   C_C, K_C, C_L, and K_L depend on the C_1/C_2 ratio and γ.

Pressure stresses and shear stresses due to V_C, V_L, and M_t are calculated by basics. Resisting areas for forces V_C and V_L are $4T(C_1$ and $C_2)$ and torsion (polar) modulus $J = C_1\ C_2(C_1^2 + C_2^2)4/3$ for Mt.

8.2.1.3 Limitations and Effects

1. β: 0.01 to 0.5 except for curves 1C, 2C: using 0.5 for higher values is conservative, Using 0.01 for lessor values will give less stresses and they are not advised
2. γ: 5 to 300, extended to 1200 in WRC 537
3. $d/D\sqrt{(D/T)} > 2$ results in unconservative error
4. The graphs are based on support at a distance of $8R_m$ each side. For $L/R_m < 8$, long plane resists deformation in circ plane at C and D; as a result, the stresses at C and D are less at more at A and B. $L/R_m = 3$ results in 10% error.
5. Radial force away from the shell results in higher stresses than it is towards the shell.
6. Rigid attachment consideration suppresses circ stresses and reverse is with long stresses
7. The stresses in thick shells are generally conservative and reverse in thin shells
8. More details are given in Appendix A2 and A3 on limitations of curves and Appendix B on K_n and K_b of WRC 107.
9. When the nozzle is relatively thin due to compensation provided in the shell, max stresses may come in the nozzle, and stresses in the nozzle shall be computed as given in 8.3.

8.2.2 Spherical Shell

Difference with WRC107: suffixes: o, i, x, and y are used in place of u, L, 2, and 1 of WRC.

Analysis for the spherical shell is as given in 8.2 and similar as for the cylindrical shell except the graphs and the following. The same analysis can be applied for elliptical heads by replacing R_m by its mean radius at junction in the applicable formulas.

1. Shell parameter U (<2.25): for round $= r_O/\sqrt{R_m/T_s}$, for square $= c/[0.875\sqrt{R_m/T_s}]$
2. Attachment parameters T' & ρ: for rigid, no parameters are required. For nozzles $T' = r_m/t$ and $\rho = T/t$, for hollow square $T' = r_m/0.875t$, $\rho = T/t$, r_m is equivalent radius with the same moment of inertia.

3. Circ and long are not relevant for the spherical shell; however, the same convention of cylinder is used to relate suffixes of forces and moments including for stresses.

Where Rm and T_S are the mean radius and thickness of the shell and r_m and t are of the nozzle.

8.2.2.1 Rigid Attachment

The resultant moment $M = M_C^2 + M_L^2$ is equivalent to separate moment for the spherical shell.

1. A *simplified procedure* is used for stresses due to P and M.
   Stress due to P, $\sigma_p = (\sigma_P C_p)(1/C_p)$, where $C_p = T^2/P$
   Stress due to M, $\sigma_m = (\sigma_m C_m)(1/C_m)$, where $C_m = T^2 \sqrt{R_m T} / M$
   Shear stress τ can be calculated as illustrated in Table 8.1
   Total stress due to pressure, P & M, $\sigma = \sigma_c + \sigma_p + \sigma_m$
   Combined stress = $P_m + P_L + Q = \sqrt{\sigma^2 + 4\tau^2} < 3S$
   Where
   - σ_c = max stress due to pressure = p R_m/T
   - The dimensionless stress parameters which can be obtained from Figure SR-1 of WRC 107 are given below.
   - $\sigma_p C_p$ = {2.6, 0.63, 0.27, 0.15, 0.093, 0.083} For U = {0.05, 0.05, 0.1, 0.15, 0.2, 0.215}
   - $\sigma_m C_m$ = {18, 1.6, 0.5, 0.205, 0.1, 0.084} For U = {0.05, 0.05, 0.1, 0.15, 0.2, 0.215}

   It is observed that axial (radial) stresses are always higher than tangential stresses.
2. *Detailed Procedure*
   Stresses in radial – Ø & tangential – r directions
   Membrane stress m = N/T = $(N/C_n)(C_n/T)$
   Bending stress b = M/Z = $6M/T^2 = (M/C_b)(6C_b/T^2)$
   Membrane + bending stress = $K_n N/T \pm K_b 6M/T^2$
   $1/C_n = T/P$, $T\sqrt{R_m T} / M$
   $1/C_b = 1/P$, $\sqrt{R_m T} / M$

The dimensionless parameters ($N_Ø$ or N_{yr})/C_n, ($M_Ø$ or Mr)/C_b in both x and y directions can be read from graphs or tables against shell parameter U from Figure SR 2 and 3 of WRC 107/537.

Calculate m & b at all eight points and then $P_m + P_L$ & $P_m + P_L + Q$ are calculated as per step 2 of 8.2.

8.2.2.2 Hollow Attachment

Stresses in radial – Ø and tangential – r directions

Membrane stress = N/T = $(N/C_n)(C_n/T)$

Bending stress = M/Z = $6M/T^2 = (M/C_b)(6C_b/T^2)$

Membrane + bending stress = K_n N/T ± K_b 6M/T^2

$1/C_n$ = T/P, $T\sqrt{R_m T}/M_C$ and $T\sqrt{R_m T}/M_L$

$1/C_b$ = 1/P, $\sqrt{R_m T}/M_C$ and $\sqrt{R_m T}/(M_L \cos\theta)$

The dimensionless parameters ($N_ø$ or N_r)/C_n, ($M_ø$ or M_r)/C_b in both x and y directions can be read from graphs or tables against shell parameter U, T, and ρ from Figure SP 1-10, SM 1-10 of WRC 107/537.

Limitations: For shells with D/t ≥ 10, d/D < 10; and with D/t ≥ 55 d/D < 1/3.

Refer Example 8.8 for spherical shell with nozzle for clarity and comparison with example in Table 8.1.

Example 8.8: For the design data of the example for the cylindrical shell in Table 8.1 calculate stresses in the spherical shell with an inside diameter of 2000mm (same as cylinder) and a thickness of 14 mm (half of cylinder) and same nozzle thickness of 10mm.

Dimensionless parameters U, T', and ρ are

R_m = mean radius of the shell = (2000 + 14)/2 = 1007

r_m = mean radius of the nozzle = (500–10)/2 = 245

U = ro/√(Rm T) = 250/√(1007*14) =2.1

T' = r_m/t = 245/10 = 24.5

ρ = T/t = 14/10 = 1.4

P_m = Stress due to pressure is the same at all eight points and in both directions

P_m = p R_m/2T = 108 MPa

Dimensionless values from graphs SP-7 for P and SM-7 for M_C and M_L are

For load P: $N_ø$/Cn = 0.01928, $M_ø/C_b$ = 0.00279, Nr/Cn = 0.5423, and Mr/C_b = 0.0051

For load M_C: $N_ø$/Cn = 0.01751, $M_ø/C_b$ = 0.00248, Nr/Cn = 0.04886, and Mr/C_b = 0.00474

For load M_L: $N_ø$/Cn = 0.01751, $M_ø/C_b$ = 0.00248, Nr/Cn = 0.04886, and Mr/C_b = 0.00474

Stresses in MPa at eight points

P_L at A, B = (N_r/Cn)(C_n/T) = 0.01928 P/T^2 ± 0.01751 $M_L/[T^2\sqrt{(R_m T)}]$ = −27.7 ± 21

Q at A, B = (M_r/C_b)(6C_b/T^2) = 0.00279(6P/T^2) ± 0.00248(6 M_L)/$[T^2\sqrt{(R_m T)}]$ = ±1 5.6 ± 12.3

P_L at C, D = (N_r/Cn)(C_n/T) = 0.01928 P/T^2 ± 0.01751 $M_C/[T^2\sqrt{(R_m T)}]$ = −27.7 ± 21

Q at C, D = (Mr/C_b)(6C_b/T^2) = 0.00279(6P/T^2) ± 0.00248(6 M_C)/$[T^2\sqrt{(R_m T)}]$ = ±15.6 ± 12.3

Shear stress τ_m = V_C/(πrm T) = ±9.1 MPa, $V_L/(\pi r_m T)$ = ±9.1 MPa

Torsional shear $\tau_2 = M_t/(2\pi r_o^2 T)$ = −1.8 MPa, $\tau = \tau_m + \tau_2$ = ±9.1−1.8 = 7.3, −10.9

Further steps are the same as example in Table 8.2 and the stresses in the shell are given below

$P_m + P_L$ = 92.2, 92.2, *112*, *112*, *112*, *112*, 92.2, 92.2, MPa, max in italics at B and C

$P_m + P_L + Q$ = 76, 107.3, 111.2, *116.4*, 111.2, *116.4*, 76, 107.3 MPa, max in italics at Bi and Ci

8.3 LOCAL STRESSES IN THE CYLINDRICAL SHELL AND NOZZLE AT THEIR JUNCTURE DUE TO EXTERNAL FORCE TENSOR ON THE NOZZLE

Local stresses in the shell are as per 8.2, similar to 8.2.1, and calculations in Table 8.1 except Cn and C_b values and geometrical constants for graphs and are given below.

Cn = Cna, Cnc, CnL = P/T, M_C/(T d), and M_L/(T d)

C_b = Cba, Cbc, VbL = P, M_C/d, and M_L/d

The dimensionless parameters ($n_ø$ or n_r) = ($N_ø$ or N_r)/C_n and ($m_ø$ or m_r) = ($M_ø$ or M_r)/C_b in both long and circ planes can be read from graphs or tables in WRC297 against the following geometrical constants of the shell and nozzle from graphs in WRC297.

T/t (<10)

d/t(10 to 100)

$\lambda = (d/D)\sqrt{(D/T)}$ (0.01 to 7).

Other limitations: nozzle inside projection and set on nozzles are not applicable. However, set on nozzles can be considered as insert nozzles without inside projection provided the shell nozzle joint is fully welded.

The detailed calculations are similar to examples in Table 8.1.

Local stresses in the shell: computation of stresses is the same as in 8.2.1 and Table 8.1 except graphs. The example in Table 8.2 gives calculations up to step 6. Step 7 is similar to the example in Table 8.1, and detailed calculations are omitted. Only the summary of stresses is shown.

Local stresses in the nozzle: circumferential membrane stresses due to P, M_C, and M_L are the same as in the shell, and bending stresses are negligible. Stresses in other directions can be better termed axial and derived below.

σ_a = membrane stress due to P = P/A, where A is the cross-sectional area of the nozzle at all eight points

σ_b = bending stress due to P = $(P/t^2)(6m_r-3n_r)$ at all eight points

σ_a = membrane stress due to M_C = M_C/Z at C and D

σ_b = bending stress due to M_C = $M_C/(t^2\ d)(6m_r-3n_r)$ at C and D

TABLE 8.2
Calculated Local Stresses in the Cyl Shell and Nozzle Due to Ext Loads on the Nozzle

1) *Input data: temp = 250°C*		2) *External loads*		3) *Input for stresses N and M*	
D = mean dia of the shell	1972	P-radial load KN	100	$\pi dT/2$	21991
T = thickness of the shell	28	VL-long load KN	100	$A = \pi[d^2 - di^2]/4$	15394
d = Nozzle OD	500	Vc-circ load KN	100	$\pi/2Xd^2T$	1.1E + 07
t = thickness of the nozzle	10	Mt-torsion moment	1E + 07	$Z = \pi[d^4 - di^4]/(32d)$	1.8E + 06
p = design pressure	3	Mc-circ moment	1E + 07	**4) *Stresses due to pressure***	
S = all stress of the shell	138	M_L-long moment	1E + 07	σc-stress = p D/2T	106
Sn = all stress of the nozzle	135	di = ID of nozzle	480	σL-stress = p D/4T	53

5) *Input parameters for graphs, Cn, and Cb required to calculate stresses due to P, MC, and ML*

		Cna	Cba	Cnc	Cbc	CnL	CbL
$\lambda = d/\sqrt{(D.T)}$	2.128	P/T	P	Mc/(T d)	Mc/d	ML/(T d)	ML/d
T/t	2.8						
d/t	50	3571	100000	714	20000	714	20000

6) *Dimensionless constants from graphs,* [Ref WRC-297]

	mr fig.no.	value	nr fig.no	value	$m_\varnothing$ fig.no	value	nØ fig.no	value
N/Cn, M/Cb for P	3 to 7	0.068	8 to 12	0.04	13 to 17	0.048	18 to 22	0.17
N/Cn, M/Cb for Mc	23 to 26	0.153	27 to 31	0.083	32 to 35	0.135	36 to 40	0.27
N/C_n, M/C_b for M_L	41 to 44	0.082	45 to 49	0.05	50 to 53	0.062	54 to 58	0.36

A) Vessel stresses: calculations similar to 7 to 10 of Table 8.1. Results are listed below.

7) *Stresses due to Fa, M_C, and M_L*; $\sigma_\varnothing$ = circumferential, σ_r = longitudinal stress

a) $\sigma_\varnothing$ circ direction	**Longitudinal plane**				**Circumferential plane**			
	Au	**A_L**	**Bu**	**B_L**	**Cu**	**C_L**	**Du**	**D_L**
Cm	74.16	74.16	92.73	92.73	98.42	98.42	102.7	102
$C = \sigma_\varnothing$	27.94	120.4	65.49	120	22.96	173.9	74.04	131
b) σ_r long direction	**Longitudinal plane**				**Circumferential plane**			
Lm	46.44	46.44	48.99	48.99	23.61	23.61	37.64	37.6
L	−18.15	111	9.5	88.48	−33.79	81.01	21.57	53.7
τ due to V_C, V_L, and M_t	**Longitudinal plane**				**Circumferential plane**			
Total shear = τ	5.46	5.46	−3.64	−3.64	−3.64	−3.64	5.46	5.46
τm = mem.shear	4.55	4.55	−4.55	−4.55	−4.55	−4.55	4.55	4.55
pm + p_L	74.9	74.9	93.2	93.2	98.7	98.7	103	103
pm + p_L + Q	47.4	123	65.7	120.4	57.2	174	74.6	131
Max of eight points	pm + p_L	103	All = 202	pm + pL + Q	174	all 405	result	OK

B) Nozzle stresses

a) $\sigma_\varnothing$ = hoop direction	**Longitudinal plane for $\sigma_\varnothing$**				**Circumferential plane for $\sigma_\varnothing$**			
1. m(P) = shell	−22.19	−22.2	−22.19	−22.19	−22.19	−22.2	−22.19	−22.2
2. m(Mc) = shell	0	0	0	0	−7	−7	7	7
3. m(ML) = shell	−9.286	−9.3	9.286	9.286	0	0	0	0
4. $\sigma_P = p(d-t)/2t$	73.5	73.5	73.5	73.5	73.5	73.5	73.5	73.5
$\sigma_\varnothing$ = sum(1 to 4)	42.02	42.02	60.59	60.59	44.29	44.29	58.32	58.3

(Continued)

TABLE 8.2 (*Continued*)
Calculated Local Stresses in the Cyl Shell and Nozzle Due to Ext Loads on the Nozzle

b) σa – axial stresses	Normal to pipe X sec, long plane				Normal to pipe X sec, circ plane			
1. membrane = P/A	−6.496	−6.5	−6.496	−6.496	−6.496	−6.5	−6.496	−6.5
2. ben = P/t²(6mr−3nr)	−288	288	−288	288	−288	288	−288	288
3. membrane = Mc/Z	0	0	0	0	−5.41	−5.41	5.41	5.41
4. b = [Mc/(t²d)] (6mr−3nr)	0	0	0	0	−133.8	133.8	133.8	−134
5. membrane = Ml/Z	−5.4	−5.4	5.4	5.4	0	0	0	0
6. b = [Ml/(t²d)] (6mr−3nr)	−68.4	68.4	68.4	−68.4	0	0	0	0
7. σ_L = p(d − t)/4t	36.75	36.75	36.75	36.75	36.75	36.75	36.75	36.8
σm = 1 + 3 + 5 + 7	24.85	24.85	35.66	35.66	24.85	24.85	35.66	35.7
σa = sum(1 to 7)	−331.6	381.2	−183.9	255.3	−397	446.6	−118.5	190
c) Shear stresses	**Longitudinal plane for τ**				**Circumferential plane for τ**			
τ1 due to Vc = 2Vc/A	13	13	−13	−13	0	0	0	0
τ2 due to Vl = 2Vl/A	0	0	0	0	−13	−13	13	13
τ3 due to Mt = Mt/(2Z)	2.7	2.7	2.7	2.7	2.7	2.7	2.7	2.7
τ = sh = $(\tau_1 + \tau_2 + \tau_3)$	15.7	15.7	−10.3	−10.3	−10.3	−10.3	15.7	15.7
τm = m.sh = $\tau_1 + \tau_2$	13	13	−13	−13	−13	−13	13	13
m_1 = σm + σø	66.87	66.87	96.25	96.25	69.14	69.14	93.98	94
$P_1 = \sqrt{[(\sigma_a - \sigma_ø)^2 + 4\tau_m^2]}$	31.15	31.15	36.01	36.01	32.45	32.45	34.48	34.5
pm + p_L	49.01	49.01	66.13	66.13	50.8	50.8	64.23	64.2
b_1 = σa + σø	−290	423	−123	316	−353	491	−60	248
$P_2 = \sqrt{[(\sigma a - \sigma ø)^2 + 4\tau^2]}$	375	341	245	196	442	403	180	135
Pm + P_L + Q	375	382	245	256	442	447	180	192

Max: pm + pl = 66, allowed = 202, pm + pl + Q = 447, allowed = 405, result = unsafe
Pm + PL = MAX(P1, ABS{m1 ± P1}/2), allowed = 1.5S: Pm + PL + Q = MAX(P2, ABS{b1 ± P2}/2), all. = 3S

σ_a = membrane stress due to $M_L = M_L/Z$ at A and B

σ_b = bending stress due to $M_L = M_L/(t^2\ d)(6m_r - 3n_r)$ at A and B

The term with n_r is introduced to correct the moment applied to the nozzle by subtracting $N_rT/2$ from the radial moment on the shell side of the junction.

The detailed calculations for stresses in the nozzle are illustrated by the example in Table 8.2 using the same data from Table 8.1.

REFERENCES

1. Code ASME S VIII D 1, 2019.
2. Welding Research council bulletin 107(537) & 297.

9 Thermal Stresses and Piping Flexibility

The introduction to thermal stresses is given in 4.4.4. Thermal stresses are of two types: one due to differential movement and other due to restriction of free movement (expansion and contraction) under temperature in operation. Fatigue life is the brittle failure on reaching maximum cycles of operation during which the element experiences bending stress from one extreme direction to the other due to repeated expansion and contraction.

In general, pressure vessels are straight between supports and rigid. Therefore, only one support is fixed and the rest are free in all directions. Even if the vessel is L shaped between fixed supports, thermal stresses will be large due to its higher diameter to length ratio (D/L) of each straight portion, and the arrangement is not practical. There can be several arrangements internal or external for pressure vessels with tubes with lower D/L, ducting, structural elements, and piping requiring restriction of free movement for process needs or to restrict vibration problems, etc., requiring flexibility analysis.

9.1 DIFFERENTIAL MOVEMENT

In boilers, pressure vessels, and heat exchangers, process requirement will force certain components to operate at different temperatures or components of different materials having different expansion coefficients are provided with a common boundary resulting in differential movement in operation. Some of the examples of different temperatures are,

- Tubes and shell of a fixed tube shell and tube boiler or heat exchanger welded to the tube sheet on both ends.
- Shell and jacket of a jacketed vessel.

An example of a component with differential materials is the butt welding of carbon and stainless steel cylindrical parts. Thermal membrane and bending stresses are induced due to the differential radial expansion and proportion to the diameter and temperature. Therefore, this arrangement is used only when the combination of size and temperature is the optimum lower limit. The stresses are solved as described in Chapter 7.

The differential temperature between the shell and tubes of a shell and tube type heat exchanger will induce stresses in the shell, tubes, and tube sheet. In boilers, the tubes containing hot gasses are immersed in water inside the shell. Due to the wetting of tubes by water, their metal temperature is marginally higher than the water

DOI: 10.1201/9781003091806-9

temperature, while the shell is at the water temperature neglecting the heat flow to the outside. Therefore, the differential movement is marginal and absorbed mainly by deflection of the tube sheet unless the length of tubes is >*6* m. In the case of other heat exchangers, if the tube sheet is rigid like the shell or the tube length is large or the shell side fluid is gas, an expansion bellow is provided in the shell to take large tube expansion. In a jacketed vessel, the jacket-closing plate ring can provide flexibility.

9.2 BASICS OF THERMAL STRESSES

Thermal stresses are direct (compression or tensile) and bending. Direct stresses in one, two, and three directions for a temperature difference ($T-T_1$) and α thermal coefficient are

$$\sigma_1 = \alpha E(T - T_1)$$
$$\sigma_2 = \alpha E(T - T_1)(1 - \nu)$$
$$\sigma_3 = \alpha E(T - T_1)(1 - \nu^2)$$

For carbon steel, at about a temperature of 100°C, direct stress σ_1 = 12.1E–6*198000*(100–21) = 189 MPa.

The above calculation indicates restriction of direct free movement, similar to the case where fixing both saddles of a horizontal pressure vessel is not safe unless the vessel operates under ambient conditions. The following arrangements of restricted free thermal movement induce bending stresses and analysis method.

- The shell or beam element is fixed at one end and the other guided end, as shown in Figure 9.1, which moves perpendicular to its axis will be in bending. Structurally, it is called a guided cantilever.

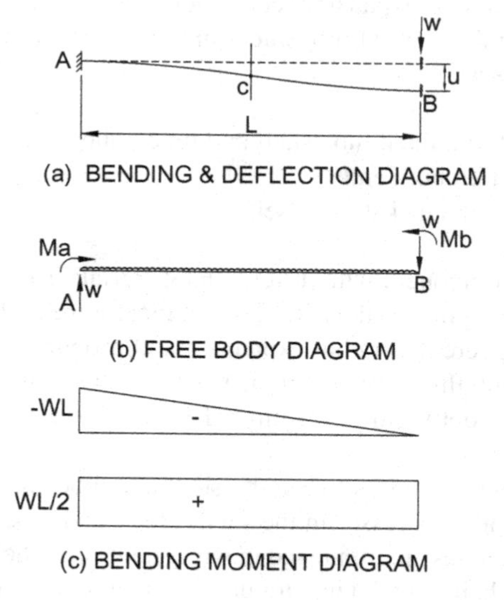

FIGURE 9.1 Guided cantilever.

- Two-element system: Elements perpendicular to each other with fixed ends deform as shown in Figure 9.2, and bending stress is induced in each element.
- Plate elements such as tube sheets and end closing plates can be analyzed by plate theory.
- 3D elements such as piping systems can be analyzed taking axisymmetric shell elements by various methods.

For any other arrangement with such thermal stresses, the components involved in the equipment or piping system between two fixed supports are stressed and generally, the most flexible is stressed the maximum. The stresses are calculated from its free body diagram which is reverse of the free body diagram in the case of loads. In this case, deflection is known instead of loads. The analysis is first to calculate forces due to deflection from the beam or relevant theory, and then stresses are calculated due to these forces by a relevant design method. The shell elements stressed due to these loads can be divided into two types, rigid elements (less flexibility or higher stiffness), which have high resistance to loads and will transfer forces and displacements without being deformed, and others that are reverse and flexible. Therefore, for calculating stresses, the general equation in a matrix form between the force and displacement is given by Eq. 9.1.

$$[F] = [S] \times [D] \tag{9.1}$$

Because equations for deflection can be conveniently derived, instead of calculating the force directly using the stiffness matrix [S], the flexibility matrix [K] is used and given by Eq. 9.2.

$$[D] = [K] \times [F] \tag{9.2}$$

After all elements of [K] are calculated by duly summing the flexibility of all components, it is inversed to convert [K] to [S] by the Excel matrix function [=MINVERSE(range)].

[F] can be solved by multiplying [D] and [S] by function [MMULT(range)]. [F] can be transferred to various points (nodes) of interest by the transformation matrix derived in 3.1.5. From forces and moments, the stresses can be compiled.

The above procedure is a basic finite element analysis for axisymmetric elements such as beam or shell elements. This procedure can be used in analysis where deflection is a factor.

9.3 GUIDED CANTILEVER

It is a beam or shell element with one end fixed (no translation and rotation) and the other end guided (no rotation and translation free), as shown in Figure 9.1. A load is applied at the guided end to give a slope or rotation of 1°. Then a moment is applied at the same point to give the same 1° rotation in the reverse direction. The net effect of the force and moment is deflection in the direction of force and no rotation. It is nothing but a pipe with one end anchored and the other end guided which moves perpendicular to its length (L) as shown in figure. To derive the equation for stress due to the expansion movement (u), two methods are given below.

(1) Divide the element (depth or pipe OD = d) into two halves; each is a cantilever with a span L/2 under concentrated load (W) at the free end. The equations for M and deflection (y) are:

$$M = W\,L/2,$$
$$y = W(L/2)^3/(3EI) \text{ or } W = 3EI\,y/(L/2)^3$$

Using basic relations $Z = I/(d/2) = 2I/d$ and bending stress $\sigma = M/Z$

$$\sigma = W\,L\frac{d}{4I} = \left[3EI\,y/(L/2)^3\right]L\frac{d}{4I} = 6E\,d\,y/L^2 \tag{9.3}$$

where E is the elastic modulus and I and Z are the moment of inertia and the section modulus, respectively.

Substituting y = u/2 in Eq. 9.3, the stress for the guided cantilever is expressed using Eq. 9.4.

$$\sigma = 6E\,d(u/2)/L^2 = 3E\,d\,u/L^2 \tag{9.4}$$

(2) Moment area method: The free body and its moment diagrams are drawn for the above pipe with W at the guided end as shown in Figure 9.1. At equilibrium $M_a = -M_b$ and taking the moment at A fixed end,

$$M_b = M_a + W\,L$$

Because $M_a = -M_b$, $M_b = W\,L/2$

Deflection at the guided end = moment of area of the moment diagram at the guided end and is given by

$$EI\,u = Mb\,L(L/2) - (W\,L^2/2)(2L/3) = (W\,L/2)L^2/2 - W\,L^3/3 = W\,L^3/12$$
$$\text{or } W = 12EI\,u/L^3$$
$$\sigma = M/Z = (W\,L/2)(d/2I) = (12EI\,u/L^3)(L\,d/4I) = d(u/2)/(L^2/2) = 3E\,d\,u/L^2$$

which is the same as Eq. 9.4,

Example 9.1: Stress using Eq. 9.4 and Figure 9.1, d = 200 mm, u = 2 mm, E = 200000 MPa, L = 1000 mm

$\sigma = 3*200000*200*2/1000^2 = 240$ MPa > allowed 138 does not satisfy.

9.4 SIMPLIFIED ANALYSIS

This analysis is mostly used for piping systems and can be used for beam and other shell elements. For beam elements, the difference is the moment of inertia.

9.4.1 3D Piping with Elements in Orthogonal Axes

General equations for stress are derived from Eq. 9.4 for three directions using the fact that the directions of u and L are perpendicular to each other. When u is in the x- direction, L direction in the 3D system is y and z. Replacing u with u_X and L^2 by $L_Y^2 + L_Z^2$ [for multiple elements $\Sigma(L_y)^2 + \Sigma(L_z)^2$],

$$\sigma_x = \frac{3E\,d\,u_x}{\Sigma L_y^2 + \Sigma L_z^2}$$

$$\sigma_y = \frac{3E\,d\,u_y}{\Sigma L_z^2 + \Sigma L_x^2}$$

$$\sigma_z = \frac{3E\,d\,u_z}{\Sigma L_x^2 + \Sigma L_y^2}$$

where u_x, u_y, and u_z are expansions of the system in three directions,

ΣL_x^2, ΣL_y^2, and ΣL_x^2 are the sum of squares of all pipe lengths in three directions. Because L is the length perpendicular to the expansion, L in E 9.4 is replaced by the sum of squares of all elements in the other two axes.

Note that $[\Sigma(L_x)]^2$ is much greater than ΣL_x^2 and a single length perpendicular to expansion is more flexible than the same total length of several pipe elements.

9.4.2 3D Piping with Elements in any of the Three Axes

Instead of three equations, 9.4.1, a single equation, 9.5, can be derived as follows:

Let u be the expansion for the distance from the anchor to the anchor AB (Figure 9.2) along its length and L_n and L_m are components of each straight pipe length ($L_a = L_1$ and L_2 in figure) in the direction normal and along the line AB, respectively.

$$L_a^2 = L_n^2 + L_m^2$$

Substituting ΣL_n^2, the sum of squares of normal components of lengths of all pipe elements for L^2 in Eq. 9.4

$\sigma = 3E\ d\ u/[\Sigma L_n^2]$, substituting $L_a^2 - L_m^2$ for L_n^2,

$$\sigma = (3Ed\ u)/\sqrt{\Sigma L_a^2 - \Sigma L_m^2} \tag{9.5}$$

Example 9.2: Calculate maximum stress at *A* (Figure 9.2) by Eq. 9.5 and, d = 0.2 m, L = L_1 = 1 m, expansion e = 0.001 m/m, E = 2e10 kg/m², and L_2 = 2 m

Angle between AB and AO, $\alpha = A\cos(L_1/L_2) = 26.6°$

Expansion of AB, $U = e\sqrt{L_1^2 + L_2^2} = 0.001*2.236 = 0.002236$ m,

$\Sigma L_a^2 = L_1^2 + L_2^2 = 1^2 + 2^2 = 5\,m^2$

$\Sigma L_m^2 = (L_1 \sin\alpha)^2 + (L_2 \cos\alpha)^2 = 0.446^2 + 1.79^2 = 3.2\,m^2$

$\sigma = 3 \times 2E11 \times 0.2 \times 0.002236\ \sqrt{(5 - 3.2)} = 18.1E7$ N/m² or 181 N/mm²

9.4.3 Code Comparison

Codes (B31.3 and IBR) specify simplified Eq. 9.6 for flexibility check of piping system between anchor to anchor without any restraints.

$$A - B > \sqrt{(d\,u/208)} \tag{9.6}$$

where A is the total developed length and B is the anchor to anchor distance in meters and d and u are in mm. Where, d = outside diameter and u = resultant expansion of the system.

Squaring Eq. 9.6, and converting units of A and B to mm

$$(A - B)^2 > 4807d\,u.$$

Converting to compare with Eq. 9.4

$$(A - B)^2 = (3E/\sigma)d\,u$$

where $(3E/\sigma) = 4807$, A–B above is the same as L in Eq. 9.4

Example 9.3: For this rule with the same data used in example 9.2, A = 3 m, B = 2.2236 m, u = 2.236 mm, d = 200 mm, A – B = 0.7764 m,

$\sqrt{d\,u/208} = \sqrt{(200*2.236/208)} = 1.46 > 0.7764$, does not satisfy.

Code Eq. 9.6 is approximately the same as Eq. 9.4 and conservative to Eq. 9.5. It is conservative for more and more elements.

Eq. 9.6 is mandatory for safe thermal stresses and if it satisfies, no further analysis is required.

9.4.4 Allowable Stress

Higher bending stress is allowed for thermal stresses than primary bending stresses as per the logic explained below.

Thermal movement is zero before commissioning and will become u under operating conditions. Assuming stress under operating conditions crosses its yield strength but still in the plastic range and within the limit of tensile strength, it yields and stress relaxation takes place and stress reduces to its yield strength. When the plant is shutdown, the reverse movement will start and first the stress value reaches zero and then it reverses and reaches yield again (in the case of design without margin). The maximum theoretical stress permitted is therefore up to twice the yield and the stress range after the first cycle is from one side yield to the other side yield. Due to this, the stress is called self-limiting. The cycle will be repeated and after reaching certain cycles depending on stress margin and material, the pipe breaks due to fatigue. The less the stress margin, the less the fatigue life. Therefore, allowable stress is limited to $1.25S_C + 0.25S_h$ in codes for piping systems and $3S$ in equipment. Here S_C and S_h are code allowed tensile stresses at ambient and maximum operating temperatures.

9.5 ANALYSIS FROM BASICS

Thermal stresses can be analyzed from basics, but practical only for a simple system without restraints. Even by using an Excel spread sheet, analysis is cumbersome for a large system with several restraints and is prone to errors.

Software such as *ceasar* and *staadpro* is designed to perform finite element analysis with axisymmetrical beam or shell elements and rigid elements (straight pipe, bend, fitting, etc.) based on the above basics.

The basics of analysis whether manual or software are explained below for a simple pipeline between two anchors (A and B) with a straight orthogonal pipe and 90° bend elements for thermal load only.

Anchor B is released, and the system is allowed to expand freely under temperature. The point B will move in three, X, Y, and Z, directions. Rotations are zero due to both ends being fixed, and represented as the deformation matrix [D]. A force tensor [F]= {F_X, F_Y, F_Z, M_X, M_Y, M_Z} is applied to bring back B to its original position.

First step: the unknown [F] at *B* is transferred to the next node and so on up to anchor point *A*. All forces are the same at all nodes, but moments will be different and include the additional moment due to the forces.

Second step: [Dn] is calculated from *A* at the next node for the first element considering a free body due to the [Fn] at that node calculated in the first step which depends on its stiffness (n is the node number). For the second element onwards, the same procedure is repeated and, in addition, rotations of the start node calculated for previous elements are multiplied by their length to obtain additional displacements.

Third step: All [Dn] are added which contain six unknowns [F] and are equated to known [D] at *B*. [F] is solved by the six equations thus obtained. [F] is obtained at all nodes by multiplying [F] at B by the transformation matrix for each element. Stresses at all nodes can be calculated from the max moment and section modulus *Z*.

This procedure is applied to solve a simple example below.

9.5.1 One-Dimensional System

Consider a beam in the X-axis of length *L* and area *A* with both ends anchored. Due to temperature, it expands by *x* when anchor *B* is freed. To bring it back, F_X has to be applied.

$$F_x = (x/L)A\,E, \text{ stress compressive} = x\,E/L = e\,E,$$

where e = expansion in m/m and equal to strain. For C&LAS at 80°C e = 0.7 mm/m, stress = 0.7*200000/1000 = 140 MPa

That is for temperatures more than 80°C, a straight pipe with both ends fixed will buckle. If the slenderness ratio increases, it will buckle at lower temperatures.

9.5.2 Two-Dimensional System

Consider one of the straight beams in *X* and *Y* axes as shown in Figure 9.2. Element *BO* is plotted in the *x*-axis and *OA* in the y-axis; L_1 and L_2 are lengths and A_1 and A_2 are areas. A and B are fixed and under temperature. *B* expands by x and y on freeing and to bring them back F_X, F_Y, and M_Z(*or M*) are applied at *B*.

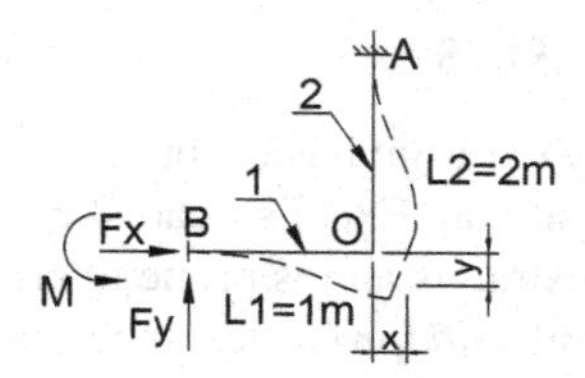

FIGURE 9.2 Two-element layout.

Step 1: Forces at O are F_X, F_Y, and $M_O = M - L_1 F_Y$

Forces at A are F_X, F_Y, and $M_a = M - L_1 F_Y + L_2 F_X$

Step 2: for element-2 (AO); displacements x_2, y_2, and θ_2 at O due to F_X, F_Y, and $M - L_1 F_Y$ are:

$x_2 = L_2^3/(3EI)F_X + (L_2^2/2)(M - L_1 F_Y)/EI$,

$y_2 = L_2/(A_2 E)F_Y$,

$\theta_2 = (L_2^2/2)F_X/EI + L_2(M - L_1 F_Y)/EI$.

For element-1 (OB); x_1, y_1, and θ_1 at B due to F_X, F_Y, and M are:

$x_1 = L_1/(A_1 E)F_X$,

$y_1 = L_1^3/(3EI)F_y - (L_1^2/2)M/EI$,

$\theta_1 = L_1 M/EI - (L_1^2/2)F_Y/EI$,

In addition, displacement at B due to θ_2, $y_{12} = \theta_2(-L_1)$, substituting the θ_2 value

$y_{12} = -(L_2^2/2)L_1F_X/EI - L_1 L_2(M - L_1 F_Y)/EI$

Neglecting direct force components x_1 and y_2 (insignificant compared to bending components) and taking the EI factor to the left side, the resultant displacements at B multiplied by EI are

$EI(x = x_1 + x_2 + x_{12}) = L_2^3/3\ F_X + (L_2^2/2)(M - L_1 F_Y)$

$EI(y = y_1 + y_2 + y_{12}) = L_1^3/3\ F_Y - (L_1^2/2)M - (L_2^2/2)L_1 F_X - L_1 L_2(M - L_1 F_Y)$

$EI(\theta = \theta_1 + \theta_2) = -(L_1^2/2)F_Y + L_1M + (L_2^2/2)F_X + L_2(M - L_1 F_Y)$

Rearranging

$$EIx = (L_2^3/3)F_X - (L_1L_2^2/2)F_Y + (L_2^2/2)M \tag{9.7}$$

$$EIy = -(L_1L_2^2/2)F_X + (L_1^3/3 + L_1^2L_2)F_Y - (L_1^2/2 + L_1L_2)M \tag{9.8}$$

$$EI\theta = (L_2^2/2)F_X - (L_1^2/2 + L_1L_2)F_Y + (L_1 + L_2)M \tag{9.9}$$

Solving these three equations, unknowns F_X, F_Y, and M can be solved.

Writing in the matrix form, EI[D] = [K][F]

$$[K] = \begin{matrix} L_2^3/3 & -L_1 L_2^2/2 & L_2^2/2 \\ -L_1 L_2^2/2 & \dfrac{L_1^3}{3} + L_1^2 L_2 & -(L_1^2/2 + L_1 L_2) \\ L_2^2/2 & -(L_1^2/2 + L_1 L_2) & L_1 L_2 \end{matrix}$$

This matrix equation can be used to calculate deflection and rotations of axisymmetric beam elements for force and moment loads.

Example 9.4: Calculate forces and max bending stress of 2D and two-element systems derived above with the same data for the pipe beam used in example 9.1 and 9.2, Figure 9.2: 200 mm OD, 7/2.2 mm thick, x-axis pipe length L_1 = 1 m, y-axis pipe length L_2=2 m, expansion about 1mm/m at 102°C, E = 2E5 MPa. Units used in calculations are: N, m Moment of inertia, I = π*0.1³*7/2200 = 1E-5, Z = 1E-4, EI x = 2E11*1E-5*1E-3 = 2E3, EIy = 4E3, θ = 0, [K] = [{2.667, −2,2},{−2,2.333, −2.5}, {2, −2.5,3}]

Solving Fx = 8250, F_Y = 27000, and M = 17000
M at other two nodes is less than that at B, max moment = 17000 Nm, bending stress = M/Z = 1.7E4/1E-4 = 1.7E8 N/m² or 170 N/mm² compared with example 9.2 (181)

9.5.3 Three-dimensional System

Manual analysis by this procedure for three or more elements or with bend elements is impractical. There are other methods (Kellogg) for solving piping systems with straight and bend elements without restraints and branches by Excel spread sheets. The flexibility matrix of each element in a layout A to B with A fixed for displacement at free end B due to the force matrix at B is derived by the above procedure and added to form the flexibility matrix [K], and a single equation in the matrix form is evolved.

$$\left[K_{6\times6}\right] X \left[F_{1\times6}\right] = EI\left[D_{1\times6}\right] \tag{9.10}$$

$$\text{Or } \left[F\right] = \left[K\right]^{-1} X\ EI\left[D\right], \text{ but } \left[K\right]^{-1} = \left[S\right] \text{ stiffness matrix and } \left[F\right] = \left[S\right] X\ EI\left[D\right] \tag{9.11}$$

Matrix Eq. 9.11 is a combination of six equations with six unknowns which can be solved easily in Excel. EI[D] is known and [K] of element depends on its geometrical properties and is derived from beam formulas. [K] is converted to [S] by Excel matrix function [=MINVERSE(range)].

[F] can be solved by multiplying [D] and [S] by function [MMULT(range)].

From [F] at B, [F] at each node point from B to A can be solved by the transformation matrix in Excel.

9.6 FORMATION OF FLEXIBILITY MATRIX OF SHELL (PIPE) ELEMENTS

Out of 6 × 6 = 36 coefficients, 15 are not required due to the symmetry of matrix [K], that is, $K_{52} = K_{25}$ and so on.

Here, 5 = row no. (r) and 2 = column no (c) in K_{52}.

K_{52} = (EI × rotation in the y direction)/Fy and K_{25} = (EI × deflection in the y direction)/My.

Out of the remaining 21, the following three elements are null elements:

Two lateral rotations due to axial moments or their reciprocal axial rotation due to lateral moments and one out of plane displacement due to axial moment or its reciprocal.

The remaining 18 elements are derived for each type of element in a 3D piping layout (straight pipe in any direction, bend of any angle, and orientation) using beam formulas. *K* for valves like elements is zero and considered as rigid. A rigid element only transfers the force but will not deform. Other fittings can be approximated as the equivalent straight element.

Pipes bends upon application of a bending load deform, and their moment of inertia is reduced by making the bend more flexible with the corresponding increase in stress. For this effect, the flexibility factor and stress intensification factors are added for bending elements to the [K] theoretically derived.

The elements of matrix [K] derived basically as displacements and rotations for axial, lateral, in plane, and out plane [D] for [F] are irrespective of the plane or axis.

Actually [K] compiled for one plane can be used for the other two planes but for different elements of [K] the following method is used.

In Krc r, c of the xy plane or (r, c)xy = (r + 1, c + 1)yz = (r + 2, c + 2)zx, in each of the three quarters:

1st = r and c number from 1 to 3

2nd = r and c number from 4 to 6

3rd = remaining r = 4 to 6 and c = 1 to 3 or 41 to 63

Example: (3, 2) in xy = (3 + 1,2 + 1 = 4, 3 but 4 jumps into the next quarter, next is back to 1, thus = 1, 3) yz = (5, 4 or 2, 1)zx.

Thus K_{32} in xy = K_{13} in YZ = K_{21} in ZX and xx = L x^2 + k $\cos^2\theta$, yy = L y^2 + k $\sin^2\theta$, xy = –(L x y + k $\sin\theta\cos\theta$), k = $L^3/12$.

Table 9.1 shows the elements of flexibility for a general straight pipe in three planes for clarity. The terms used in this table are defined in Table 9.2 and Fig. 9.3.

TABLE 9.1
Elements of [K] for 3D Pipe

K_{XX}	YZ-plane	ZX	XY
11	a.b	xx + z^2b	yy + z^2a
22	yy + z^2a	a.b	xx + z^2b
33	xx + z^2b	yy + z^2a	a.b
21	–z.ao	–z.bo	–xy – z^2q
32	–xy – z^2q	–z.ao	–z.bo
13	–z.bo	–xy – z^2q	–z.ao
41	0	x.q	–z.q
52	–z.q	0	x.q
63	x.q	–z.q	0
51	–bo	–L.x	–z.a
62	–z.a	–bo	–L.x
43	–L.x	–z.a	–bo
61	ao	z.b	L.y
42	L.y	ao	z.b
53	z.b	L.y	ao
44	L	a	b
55	b	L	a
66	a	b	L
65	q	0	0
46	0	q	0
54	0	0	q

EI being constant for the same size of element under consideration, suppressed in equations hereafter.

9.6.1 Single Element in Plane Forces

For a straight pipe *ab* of length *L* with fixed *a*, the force tensor [F] is applied at *b* (Figure 9.3), and displacement tensor [D] at *b* as per beam formulas is compiled below (elements indicated are for the X direction pipe in the XY plane).

Lateral displacement due to lateral forces (K_{22}, K_{33}) = $L^3/3$

Lateral rotation due to lateral force = displacement due to lateral moments (K_{26}, K_{35}) = $L^2/2$

Lateral rotation due to lateral moments (K_{55}, K_{66}) = L

Axial rotation due to axial moment L/(G J), K_{44} = 1.3L

where G = rigidity modulus = E/[2(1 + v)] = E/2.6 for steels (v = 0.3)

J = polar moment of inertia = 2I for shell elements, I = MI

Effects of direct and shear stresses like displacement due to axial force are neglected compared to that due to lateral loads and moments.

Using the above formulas, all the elements of [K] can be computed.

For inclined pipes in any plane, the equations for elements are derived by resolving forces axial and lateral to the pipe, the [D] in the rotated angle is calculated, and then [D] is resolved back in to orthogonal axes. The above elements are tabulated in column one of Table 9.2.

9.6.2 3D System

For 3D piping layout, all out-of-plane forces in addition induce bending and/or torsion moment. The actual layout of *A* to *B* shown in Figure 9.3 involves several elements in all planes; the method is to calculate the *k* that is [D] at *B* due to [F] at *B* for each pipe element considering that all other elements are rigid. *k* of all elements of each plane are added separately and thus added elements are added in different combination (see ref. 1 for details of combination) to form [K].

[K] can be calculated by programming using the following procedure using matrix equations.

The force tensor at *B* can be transferred to *b* and rotated to align with pipe element *ab* (X - axis in Fig. 9.3) and the matrix *K* can be computed as per 3.1.5 and 3.1.6 using T and R from Table 3.1, as follows.

$[F_B]_{61}$, $[Fb]_{61}$, and $[F]_{61}$ are the force matrix at B, b, and (b.local)

$[DF_B]_{61}$, $[Db]_{61}$, and $[D]_{61}$ are the displacement matrix at B, b, and (b.local)

$[T_f]_{66}$ = transformation matrix for F from B to b

$[R_f]_{66}$ = rotation matrix at b rotating relevant axis to align with the element (Z - axis for Fig. 9.3) for F

$$[F]_{61} = [R_f]_{66}[Fb]_{61} = [R_f]_{66}[T_f]_{66}[F_B]_{61} \tag{9.12}$$

If the [R] at *b* is different from the one given in Table 3.1, it can be computed based on the local direction cosines.

k is for *d* at *b* for force *F*, and the elements of k are given in Table 9.2 (column one for x - axis).

Similarly, the local matrix D at b can be rotated and transferred back to B and given by the equation.

$$\left[D\right]_{61} = \left[T_d\right]_{66}\left[R_d\right]_{66}\left[D_B\right]_{61} \tag{9.13}$$

Substituting, [D] and [F] from Eq. 9.12 and 9.13, in equation [D] = [K] [F] at point b in local axes

$$\begin{aligned}
\left[T_d\right]_{66}\left[R_d\right]_{66}\left[D_B\right]_{61} &= \left[K\right]\left[R_f\right]_{66}\left[T_f\right]_{66}\left[F_B\right]_{61} \\
\left[D_B\right] &= \left\{\left[R_f\right]\left[T_f\right]/\left[T_d\right]\left[R_d\right]\right\}\left[k\right]\left[F_B\right] = \left[K\right]\left[F_B\right] \\
\left[K\right] &= \left[k\right]\left[R_f\right]\left[T_f\right]/\left[T_d\right]\left[R_d\right] \\
\left[K\right] &= \left[T_d\right]^{-1}\left[R_d\right]^{-1}\left[k\right]\left[R_f\right]\left[T_f\right]
\end{aligned} \tag{9.14}$$

[K] can be derived analytically for one general straight pipe element by beam theory or using Castigliano's first theorem.

9.6.2.1 Beam Theory

Consider one element ab (a-fixed, b-free) in the x-axis and XY plane at a distance (X, y, z) from b to B as shown in Figure 9.3.

Deflection y of the pipe element of length L for in plane lateral force F_Y (unit F_Y EI) at any point B at a distance x and y from b of element connected through rigid elements as shown in Figure 9.3 is derived as:

F_Y is transferred to b as F_Y and moment $F_Y X$

Deflection and rotation due to F_Y at b on ab are $d_1 = Fy\, L^3/3$ and $\theta_1 = F_Y L^2/2$.

Deflection and rotation due to moment $F_Y X$ at b on ab are $d_2 = F_Y X L^2/2$ & $\theta_2 = F_Y X L$

Deflection at B due to rotations θ_1 and θ_2 is

$$d_3 + d_4 = (\theta_1 + \theta_2)X = X(L^2 F_Y/2 + L F_Y X) = (L^2 X/2 + L X^2)F_Y$$

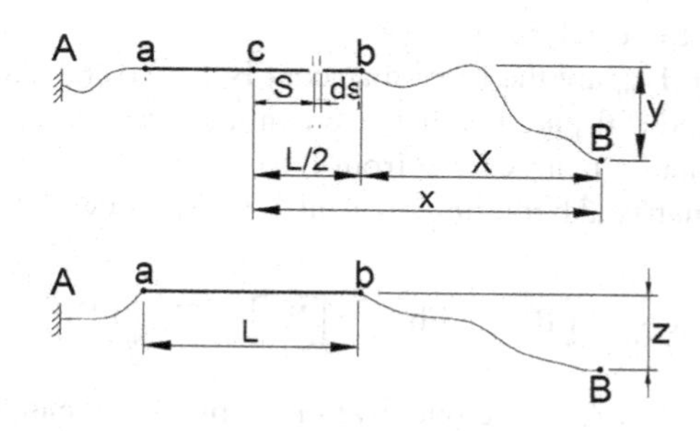

FIGURE 9.3 3D layout of a straight pipe element.

Total deflection at B, $K_{22} = d_1 + d_2 + d_3 + d_4 = (L^3/3 + L^2 X/2 + L^2X/2 + L X^2) F_Y$, writing $L^3/3 = L^3/12 + L^3/4$

$$K_{22} = L^3/12 + L^3/4 + L^2X + L X^2$$
$$K_{22} = L^3/12 + L\left(L^2/4 + L X + X^2\right) \quad (9.15)$$
$$= [L^3/12 + L\left(L/2 + X\right)^2$$

Because L/2 + X = x component of the distance from B to the mid-point (C) of ab say x as shown in Figure 9.3, and substituting in Eq. 9.15, K_{22} is simplified to Eq. 9.16

$$K_{22} = L x^2 + L^3 / 12 \quad (9.16)$$

9.6.2.2 Castigliano's First Theorem

Castigliano's first theorem states that the partial derivative of the strain energy with respect to any one of the static loads is equal to the displacement of the point of application of that load in the direction of force. The above Eq. 9.16 can be derived by Castigliano's first theorem.

Take element *ds* at distance *s* from midpoint C as shown in Figure 9.3. Moment at *ds* for in plane forces: Fx, Fy, and Mz at point B

$$M = (x - s) F_Y + y F_X + Mz$$

Strain energy at d_S, $U = M^2 ds/(2EI)$
As per Castigliano's first theorem, displacement d_Y at the point of force = dU/dF_Y
Substituting the *M* value for *U* in the *dy* equation, *dy* is expressed as

$$d_Y = d/dF_Y[(x - s) F_Y + y F_X + Mz]^2 d_S/(2EI)$$

Suppressing EI, and differentiating with respect to dF_Y

$$d_Y = [(x - s) F_Y + y F_X + Mz](x - s)ds$$
$$d_y = \int \left[(x - s)^2 F_Y + y(x - s)F_X + M_Z(x - s)\right] ds$$
$$= \int_{L/2}^{-L/2} [(x^2 + s^2 - 2xs)\left[F_Y + y(x - s)F_X + M_Z(x - s)\right] ds$$

Integrating, and applying limits

$$d_Y = \left[(x^2s + s^3/3 - 2 \times s^2/2) F_Y + y(x\, s - s^2/2)F_X + (x\, s - s^2/2)\, M_Z\right]$$
$$d_Y = (x^2L + L^3/12)F_Y + y \times L\, F_X + x\, L\, M_Z = K_{22}F_Y + K_{21}F_X + K_{26}M_Z$$

$K_{22} = x^2 L + L^3/12$ is the same as Eq. 9.16

That is deflection at B in the y direction due to F_Y at B is given by Eq. 9.16. If the ab is inclined at an angle θ anticlockwise from the X axis, the component of force F_Y lateral to ab = $F_Y\cos\theta$. The deflection $L^3/12$ will be = $\cos\theta\, L^3/12$ lateral to ab. To obtain deflection in the Y direction which is the component of $\cos\theta L^3/12$, that is $\cos^2\theta\, L^3/12$ (K_{22}) for the inclined element, is given by

$$K_{22} = L\, x^2 + \cos^2\theta\, L^3/12 \tag{9.17}$$

Similarly, $\cos\theta\cos\theta\, L^3/12$ is added to K_{21}. No addition to K26 as M is in the Z axis normal to plane XY under consideration.

In a 3D layout as the F_Y as the out-of-plane force will induce torsion moment ($F_y\, z$) which rotates element and additional deflection of point B and can be derived by the above theory and given by

$$L\, z^2(\sin^2\theta + 1.3\cos^2\theta).$$

$$K_{22} = L\, x^2 + \cos^2\theta\, L^3/12 + L\, z^2(\sin^2\theta + 1.3\cos^2\theta) \tag{9.18}$$

Similarly, $(z^2 0.3L\cos\theta\sin\theta)$ is added to K_{21}.

All elements can be derived by this theorem using the above procedure as follows:

1. Compute moments (1-axial out of plane, 2-lateral out of plane, 3-lateral in plane) at C due to relevant forces/moments at B

   M1 = km Mx + km My + kn Fy + kn Fz

   M2 = km Mx + km My + kn Fx + kn Fz

   M3 = km Mz + kn Fx + kn Fy

   Km and kn are expressions each depending on x, y, z, L, and θ
2. Derive equations [D] at B using the above theorem from the above equations each by differentiating with respect to each force forming 4+4+3 equations.
3. Compute K from each equation. Seven of the expressions (11, 22, 33, 44, 55, 45, and 53) are found in the two equations which are added to obtain total 18 elements forming the required [K].

Table 9.2 is shown with six columns with a straight pipe element in different positions with one end (a) fixed and the other end (b) applied force in the first two columns, rest connected to B through assumed rigid elements and applied forces at B as shown in Figure 9.3. Expressions in the first four columns and some in the last two columns can be derived by beam theory. The rest are complicated and can be derived by the above procedure. For pipe ab not in any plane formation of moments, M1, M2, and M3 are complicated, and [K] can be derived by the matrix method explained in 9.6.2.

For convenience to program in Excel, the load and deflection point can be taken at any convenient point within the layout and the origin (0,0,0) of the coordinate system can be considered. Forces and deflections are solved at the origin and can be transferred to any node point.

TABLE 9.2
Elements of Flexibility Matrix for a Straight Pipe in the XY Plane

	X, E	Y, E	X, P	Y, P	XY, P	XY, A
	1	2	3	4	5	6
11	0	$L^3/3$	$L.y^2$	$L^3/3$	$L.y^2 + k\sin^2\theta$	$5 + z^2a$
22	$L^3/3$	0	$L^3/3$	$L.x^2$	$L.x^2 + k\cos^2\theta$	$5 + z^2b$
33	$L^3/3$	$L^3/3$	$1.3L^2$	$1.3L^2$	a.b	5
44	1.3L	L	1.3L	L	b	5
55	L	1.3L	L	1.3L	a	5
66	L	L	L	L	L	5
12	0	0	–L.x.y	–L.x.y	$4–K\cos\theta\sin\theta$	$5 + q.z^2$
13	0	0	0	0	0	–z.ao
14	0	0	0	0	0	–q.z
15	0	0	0	0	0	–z a
16	0	$L^2/2$	L.y	L.y	4	5
23	0	0	0	0	0	–z.bo
24	0	0	0	0	0	z.b
25	0	0	0	0	x.q	5
26	$–L^2/2$	0	–L.x	–L.x	4	5
34	0	$–L^2/2$	–1.3L.y	–y.L	4	5
35	$L^2/2$	0	L.x	1.3L.x	x.a – y.q	5
36	0	0	0	0	0	0
45	0	0	0	0	q	5
46	0	0	0	0	0	0
56	0	0	0	0	0	0

Pipe oriented along (X-x axis, Y – y axis, XY – inclined); application of force and displacement at (E – end point, P- any point in-plane, A – any point x,y,z)
$C = \cos\theta$, $s = \sin\theta$, $k = L^3/12$, $q = 0.3L.c.s$, $a = L(c^2 + 1.3s^2)$, $b = L(s^2 + 1.3C^2)$,
$ao = x.a – y.q$, $bo = x.a – y.q$, $a.b = x.ao + y.bo + k$

Castigliano's theorem can be applied individually to derive each element instead of combining as above.

9.7 FORMATION OF FLEXIBILITY MATRIX FOR BEND ELEMENTS

Deformation and rotations of bend pipe in orthogonal planes due to external loads as shown in Figure 9.4 are derived based on curved beam theory and Castigliano's first theorem below, for nonorthogonal oriented by the matrix method explained in section 9.6.2.

Refer Figure 9.4 showing one bend element (a-b) having coordinates (x, y, and z) at center (o) of the bend in the XY plane and orientation angle α, with any convenient origin say B in the 3D layout (A-B) with either side elements considering rigid, under force matrix [F] applied at origin *B*. For simplicity, first consider the 2D layout with $z = 0$, bend angle $\theta = 90°$ and $\alpha = 0$ then add effects of z, any bend angle θ, and any orientation angle α.

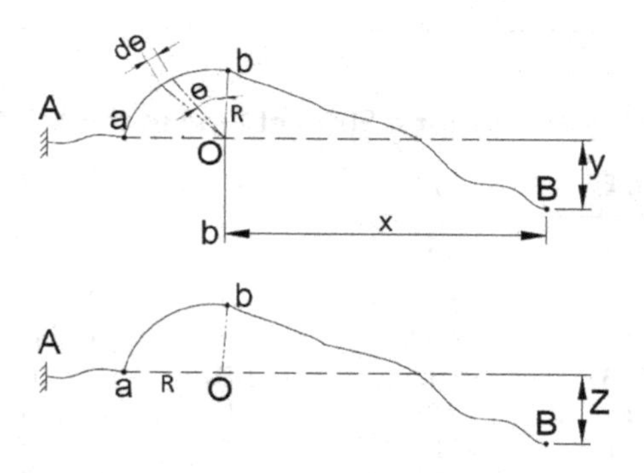

FIGURE 9.4 3D layout of the bend pipe element.

Here

θ = bend angle + when rotated anticlockwise from start point *b*, - when rotated clockwise.

α = angle from negative vertical axis to bend radius at start point *b* A'clock wise.

Consider infinitesimal length ds (dθ) at an *angle θ* (*not bend angle and only used in integration*) from radius *o-b* as shown in Figure 9.4. In plane moment at ds due to in plane forces F_X, F_Y, and M_Z is given by

$$M = F_X(y + R\cos\theta) + Fy(x + R\sin\theta) + M_Z$$

Strain energy on *ds*, $U = M^2\, ds/(2EI)$

By Castigliano's first theorem, deformation in the x direction dx is given by the equation:

$$dx = dU/dF_X,\ dy = dU/dF_Y \text{ and } \theta z = dU/dM_Z$$

$dx = d/dF_X[M^2\, ds/(2EI)]$, suppressing EI, substituting M and ds = R dθ and partial differentiating

$dx = [F_X(y+R\cos\theta)+F_Y(x+R\sin\theta)+M_Z]R\, d\theta$

$dx = [F_X(y+R\cos\theta)^2+F_Y(x+R\sin\theta)(y+R\cos\theta)+M_Z(y+R\cos\theta)]R.d\theta$

$dx = \{F_X(y^2+R^2\cos^2\theta+2yR\cos\theta)^2+F_Y[x\, y+R^2\sin\theta\cos\theta+R(x\cos\theta+y\sin\theta)] +M_Z(y+R\cos\theta)\}R\, d\theta$

Integrating with limits 0 to $\pi/2$

$dx = \{[y^2R\theta+R^3(\theta/2+\sin\theta\cos\theta+2yR^2\sin\theta)\, F_X + [xyR\theta+R^3\sin^2\theta/2+R^2 (x\sin\theta-y\cos\theta)]F_Y+[yR\theta+R^2\sin\theta)\, M_Z\}$ 0 to $\pi/2$

$dx = (y^2\, R\pi/2+R^3\, \pi/4+2yR^2)\, F_X+(x\, y\, R\pi/2+R^3/2+R^2x+R^2y) F_Y+(y\, R\, \pi/2+R^2)\, M_Z=K_{11}\, F_X+K_{12}\, F_Y+K_{16}\, M_Z$

Similarly, partial differentiating U with respect to F_Y and suppressing F_X as it gives $K_{21} = K_{12}$

$dy = Fy(x+R\sin\theta)^2 + Mz(x+R\sin\theta)]R.d\theta$, integrating and applying limits

$= (x^2R\pi/2 + R^3\pi/4 + 2xR^2)Fy + (xR\pi/2 + R^2)M_Z = K_{22}.F_Y + K_{26}.M_Z$

Similarly, partial differentiating U, integrating, and applying limits with respect to M_Z and suppressing F_X and F_Y as it gives K61 = K16 and K62 = K26

$$\theta_Z = R\pi/2M_Z = K66.M_Z$$

Derivation of other coefficients of matrix [K] for out-of-plane moments due to F_Z, M_X, and M_Y at ds is calculated as follows:

There are two moments at ds, one axial (Ma) and other transverse (Mt).

$M_a = F_Z(x + R\sin\theta)\cos\theta + F_Z(y + R\cos\theta)\sin\theta + M_X\cos\theta + M_Y\sin\theta$

$M_t = F_Z(x + R\sin\theta)\sin\theta + F_Z(y + R\cos\theta)\cos\theta + Mx\sin\theta + M_Y\cos\theta$

$U_a = M_a^2Rd\theta/2$, $Ut = M_t^2Rd\theta/2$

Due to M_a:

$\theta_{xa} = dU_a/dM_X = 1.3/k\{F_Z[(x+R\sin\theta)(\cos^2\theta+\sin\theta\cos\theta)] + M_X\cos^2\theta - M_Y\sin\theta\cos\theta\}Rd\theta$

$\theta_{ya} = dUa/dMy = 1.3/k\{F_Z[(x+R\sin\theta)(\sin^2\theta+\sin\theta\cos\theta)] + M_X\sin\theta\cos\theta + M_Y\sin^2\theta\}Rd\theta$

1.3/k is due to torsion (1/GJ = 1.3/EI) and *flexibility factor k* is added because all coefficients of bend are multiplied by k Q and *k* is not applicable for torsion.

Due to Mt:

$\theta_{xt} = dU_t/dM_X = \{F_Z[(x+R\sin\theta)(\sin^2\theta+\sin\theta\cos\theta)] + M_X\sin^2\theta - My\sin\theta\cos\theta\}R\,d\theta$

$\theta_{yt} = dUt/dM_Y = \{F_Z[(x+R\sin\theta)(\cos^2\theta+\sin\theta\cos\theta)] + M_X\sin\theta\cos\theta + M_Y\cos^2\theta\}R\,d\theta$

Omitting F_Z portion (too long and complicated)

$\theta_X = \theta_{xa} + \theta_{xt} = (1.3/k\cos^2\theta + \sin^2\theta)M_X R\,d\theta + (\sin\theta\cos\theta - 1.3/k\sin\theta\cos\theta)M_YR\,d\theta$

Integrating

$\theta_X = (1.3/k\ \ \theta/2 + 1.3/k\sin\theta\cos\theta) + \theta/2 - \sin\theta\cos\theta/2)R\ \ M_X + (\sin^2\theta/2 - 1.3/k\ \ 0.5\sin^2\theta)R\,M_Y$

Applying limits

$\theta_X = (1.3/k\ \pi/4 + \pi/4)R\,M_X + (1/2 - 2.6/k)R\,M_Y$

$= K_{44}\,M_X + K_{45}\,M_Y$

Similarly suppressing Mx portion as it gives $K_{54} = K_{45}$

$$\theta_Y = \theta_{ya} + \theta_{yt} = (1.3/k \sin^2\theta + \cos^2\theta) M_Y\, R\, d\theta$$

Integrating $\theta_Y = (1.3/k\ \theta/2 - 1.3/k \sin\theta\cos\theta) + \theta/2 + \sin\theta\cos\theta) R\, M_Y$

Applying limits $\theta_Y = (1.3/k\ \pi/4 + \pi/4) R\, M_Y = K_{55}\, M_{YY}$

Similarly, the last three coefficients K_{33}, K_{34} and K_{35} can be derived from dU/dF_Z and M_a and M_t (derivation omitted) and are tabulated in the middle column of Table 9.3. Remaining 6 out of 18 are zero for this data.

TABLE 9.3
Elements of Flexibility Matrix for a Pipe Bend in Various Positions in General Piping Layout

α = angle of tangent at start point of bend from positive horizontal axis

θ = bend angle, + when rotated anticlockwise from start point, – clockwise
x, y, z are coordinates of center of bend radius with respect to the load point

Cx = cos α – cos (α + θ) L = Rθ		a = (Cyy + 1.3Cxx/k)R
Cy = sin α – sin (α + θ)		b = (Cxx + 1.3Cyy/k)R
Cxx = 0.5θ – 0.25[sin2(α + θ) – sin2α]		ao = x.a – y.q + 1.3Cx.R²/k
Cyy = 0.5θ + 0.25[sin2(α + θ) – sin2α]		bo = y.b – x.q + 1.3Cy.R²/k
Cxy = 0.25[cos2(α + θ) – cos2α]		xx = Cxx.R³ + 2Cx.x.R² + L.x²
xy = Cxy.R³ + (Cx.y + Cy.x)R² + L.x.y		yy = Cyy.R³ + 2Cy.y.R² + L.y²
ab = 2.6(Cx.x + Cy.y + R.θ/2)R²/k + x²a + y²b + 2x.y.q		q = (1 – 1.3/k)Cxy.R

	3D XY	**2D XY:** α = 0, θ = 90°, L = πR/2	*Cx = Cy* = ±1, *Cxx = Cyy* = π/4, Cxy = 1/2
11	yy + z²a	y²Rπ/2 + R³ π/4 + 2y.R²)	Cyy.R³ + 2y.y.R² + L.y² = yy
21	–xy – z²q	–(x.y Rπ/2 + R³/2 + R²x + R²y)	Cxy.R³ + (Cx.y + Cy.x)R² + L.x.y = xy
31	–z.ao	0	0
22	xx + z²b	x²R.π/2 + R³π/4 + 2x.R²)	Cxx.R³ + 2Cx.x.R² + L.x² = xx
32	–z.bo	0	0
33	a.b	K44.K55	a.b
41	–z.q	0	0
51	–z.a	0	0
61	L.y + Cy.R²	y.R π/2 + R²	L.y + Cy.R²
42	z.b	0	0
52	x.q	0 for θ = 90°	0
62	–L.x – Cx.R²	–(x.Rπ/2 + R²)	–(L.x + Cx.R²)
43	–bo.θ	–(y.K44 – x.q ± 1.3R²)	y.b – x.q ± 1.3Cy.R/k = bo
53	ao	x.K55 – y.q ± 1.3R²	x.a – y.q ± 1.3Cx.R²/k = ao
63	0	0	0
44	b	(1.3/k π/4 + π/4)R	(Cxx + 1.3Cyy/k)R = b
54	q	R/2(1 – 1.3/k)	Cxy.R(1–1.3/k) = q
64	0	0	0
55	a	(1.3/k π/4 + π/4)R	(Cyy + 1.3Cxx/k)R = a
65	0	0	0
66	Rθ	R π/2	L

All elements are multiplied by k.Q, k-flex. factor, and Q stress int. factor; k is not applicable for torsion; hence, torsion effects are corrected by 1.3/k for 1.3

For any angle of α and bend angle, coefficients as trigonometric functions of α and θ are added. The coefficients Cx, Cy, Cxx, Cyy, and Cxy for which 1 or $\pi/4$ or 1/2 for $\theta = 90°$ and $\alpha = 0$ are part of the above derivations. The equations for coefficients are given in the table, and the elements of [K] are listed in the last column of table adding the above coefficients replacing their constant values given in the middle column for the above analysis.

For a 3D system ($z \neq 0$), all additions due to the z coordinate effect are added to the elements of the last column and listed in the first column of the table which gives expressions for all elements of [K] for the general bend element (bent pipe) for the XY plane. For the other two planes, the elements can be listed as per 9.6 and Table 9.1.

REFERENCE

1. Kellogg handbook.

10 Flat-Plate Components

Flat plates are extensively used in various shapes as pressure or structural parts in pressure vessels. Pressure components using flat plates are mostly:

1. End closures of cylindrical or rectangular shells: These are either welded or flanged. Flanges are covered in Chapter 13.
2. Rectangular air or gas ducts: These are used when the size is large and pressure is low as an alternative to the cylindrical shape.
3. For enclosing rectangular heat exchangers and end closures of jacketed vessels.
4. Tube sheet (TS) of the shell and tube heat exchangers.

The flat plate component is simple but requires higher thickness and more weight for the pressure part compared to cylindrical, spherical, and other formed parts, practically economical only to small sizes and low pressures.

For larger sizes and higher pressures, thickness can be reduced by providing stiffeners.

10.1 FLAT PLATE THEORY

Notation

a = outer radius of the circular plate or larger side of the rectangular plate

b = inner radius of the circular plate or smaller side of the rectangular plate

E = elastic modulus, E_e = plate equivalent of E due to biplane stresses = $E/(1 - v^2)$

I = moment of inertia per unit width (circumference) = $t^3/12$

D = plate constant = $E_e I = E.t^3/[12(1 - v^2)]$, (unit-Nmm)

M = moment per unit width (circumference)

n = b/a ($n \leq 1$)

P = pressure

t = thickness

y = deflection

θ = slope or rotation

v = Poisson's ratio

SS = simply supported

Subscripts: c = center, e = edge, r = radial, t = tangential

DOI: 10.1201/9781003091806-10

Unlike the beam element (axi symmetric), the plate element is bi plane. Poisson's ratio v is factor in plate parameters, as in Ee indicated in notation. Stresses are largely of bending with a small amount being membrane type. Analysis of flat plate theory is not valid for very thick and very thin plates. Thickness should be

- Less than about one quarter of the least size. t should be <100/4 for the plate of size 100 × 200.
- Minimum not to deflect more than about one-half the thickness.

A flat plate can be any shape, but in pressure vessels, for the purpose of analysis, three shapes are normally found, circular solid, circular ring, and rectangular. For economy, stiffeners are part of the design unless the size is very small.

Deflection, slope, moment, and shear force at any point can be derived by the integration method (Ref. 2). For any load and boundary condition, two moments are induced perpendicular to each other at any point. To understand, consider a circular ring of uniform thickness SS at the entire outer edge and free and line load at the inner edge. Let a radial strip of sufficient width compared to its thickness be considered, which in isolation is unstable, and rotated at the outer edge due to radial moment without any resistance. However, the ring is stable and will not rotate freely, because of the tangential moment induced right angle to the radial moment to resist free rotation. Shear force and moment are the reactions at the cutout edges of the strip in its free body diagram.

General equations for moments M and deflection y of a flat plate of constant thickness t under pressure P can be derived and given by

$$M = k_m P a^2 \tag{10.1}$$

$$y = k_y (P/E) a^4 / t^3 \tag{10.2}$$

where k_m and k_y are constants that depend on Poisson's ratio v and n and on the condition and shape of the boundary. The maximum moment as can be observed in the following sections is generally the same as that for the beam at the edge for fixed and at the center for SS.

10.2 CIRCULAR PLATES

A general equation for moments per unit circumference in radial and tangential directions at any point on any radius (r) on the circular plate without the boundary condition and outside radius can be derived by the integration method by the theory of elasticity under pressure P as shown in Figure 10.1 which shows a radial strip of angle $d\alpha$ and dr with radial and tangential moments. The general equations thus derived are given by

$$y = y_C + \frac{M_C r^2}{2D(1+v)} - \frac{P r^4}{16D} \tag{10.3}$$

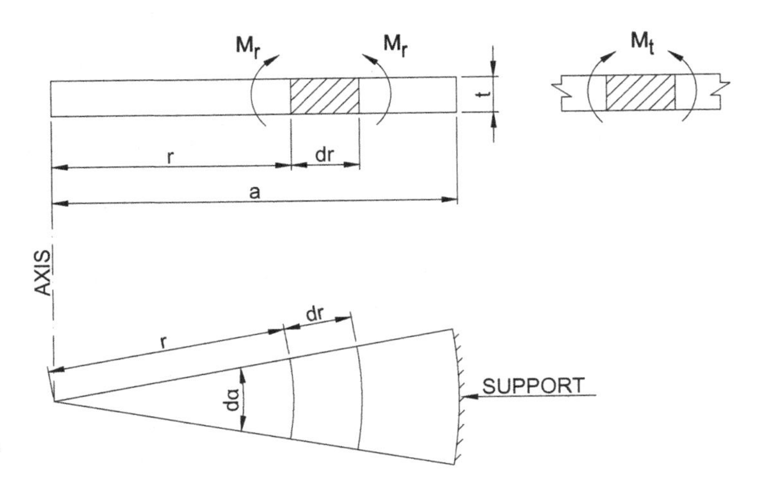

FIGURE 10.1 Circular plate.

$$\theta = \frac{M_C r}{D(1+v)} - \frac{P r^3}{16D} \tag{10.4}$$

$$M_r = M_C - P r^2 (3+v)/16 \tag{10.5}$$

$$M_t = \theta D(1-v^2)/r + v M_r \tag{10.6}$$

Substituting the expression for θ and M_r from Eqs. 10.4 and 10.5 in 10.6

$$M_t = \left[M_C/(1+v) - Pr^2/16\right]\left(1-v^2\right) + v\left[M_C - P^2 r^2 (3+v)/16\right]$$

Simplifying

$$M_t = M_C - \frac{P r^2}{16}(1+3v) \tag{10.7}$$

where M_c and y_c are the moment and deflection at the center and can be derived with known boundary conditions.

10.2.1 Simply Supported

M_C can be derived by integration and given by Eq. 10.8

$$M_C = (3+v) P a^2/16 \tag{10.8}$$

Substituting Eq. 10.8 in Eqs. 10.5 and 10.7, Eqs. 10.9 and 10.10 are obtained

$$M_r = \left[(3+v)\left(a^2 - r^2\right)\right]\frac{P}{16} \tag{10.9}$$

$$M_t = \left[a^2(3+v) - r^2(1+3v)\right]\frac{P}{16} \tag{10.10}$$

M_r is similar to the beam maximum at the center and reduces to zero at the edge.

M_t is the maximum at the center and reduces to a minimum at the edge.

Substituting, zero for r in Eqs. 10.9 and 10.10, moments at the center M_{rc} and M_{tc} and a for r moments at the edge M_{re} and M_{te} are obtained.

$$M_{rc} = M_{tc} = (3+v)P\,a^2/16 \tag{10.11}$$

$$M_{re} = 0 \quad \text{and} \quad M_{te} = (1-v)P\,a^2/8 \tag{10.12}$$

From above Eqs. 10.9 to 10.12, the maximum moment is at the center.

Substituting $v = 0.3$, the maximum moment (at the center) is $M = 0.2065P\,a^2$ (k_m in Eq. 10.1 = 0.2065)

Corresponding stress = $M/z = 6M/t^2$

$$\sigma = 1.2P(a/t)^2 \tag{10.13}$$

For ($P = 1$ MPa, $a = 500$ mm, and $t = 50$ mm), $\sigma = 120$ MPa.

10.2.2 Clamped (Fixed)

Moment at the center can be derived as

$$M_C = (1+v)P\,a^2/16 \tag{10.14}$$

Substituting Eq. 10.14 in Eqs. 10.5 and 10.7, Eqs. 10.15 and 10.16 are obtained

$$M_r = \left[a^2(1+v) - r^2(3+v)\right]P/16 \tag{10.15}$$

$$M_t = \left[a^2(1+v) - r^2(1+3v)\right]P/16 \tag{10.16}$$

Substituting, zero for r in Eqs. 10.15 and 10.16, moments at the center M_{rc} and M_{tc} and a for r moments at the edge M_{re} and M_{te} are obtained.

$$M_{rc} = M_{tc} = a^2(1+v)P/16$$
$$M_{re} = -Pa^2/8, M_{re} > M_{rc}$$
$$M_{te} = -vPa^2/8, M_{tc} > M_{te}$$

M_r and M_t reduce to zero between the center and edge and reverse and increase for the maximum at the edge. The maximum moment from the above equations is *Mre* at the edge and is given by Eq. 10.17

$$M = P\,a^2/8(k_m = 1/8) \tag{10.17}$$

and

$$\text{Maximum stress} = M/z = 6M/t^2 = 0.75P(a/t)^2 \tag{10.18}$$

For (P = 1 MPa, a = 500 mm, and t = 50 mm), σ = 75 MPa.

Practically boundary conditions are neither SS nor clamped. The joint between the shell and flat-plate end closure will rotate, when the clamp is released and the moment at the joint is redistributed between the end plate and shell depending on their stiffness. The moment in the plate will reduce and induce a moment in the shell. The effect of this rotation will increase the moment at the center. However, a higher stress is allowed due to cyclic loading as relaxation takes place and stress will reduce at the cost of fatigue. The value of k_s in general Eq. 10.1, $k_sP(a/t)^2$ for maximum stress at the center of the plate, varies from 0.5 to 1.25 depending on the type of plate joint with the connecting part. K_S = 0.75 for fixed, 10/8 or 1.25 for SS with v = 0.3.

The code[1] gives the equation t = D√(C.P/S) or σ = CP(D/t)² where D = outside diameter

The C value depends on the type of shell to plate joint and is equal to 0.3 for flanged, 0.33 for welded (nearer to SS), and 0.17 for integral (nearer to fixed).

For (P = 1 MPa, a = D/2 = 500 mm, and t = 50), σ for SS = 132MPa and for fixed = 68MPa almost the same as (120 by Eq. 10.13) and (75 by Eq. 10.18).

Moments and stresses in circular plates or rings for pressure and other loads can be derived by the integration method. Ref. 2 gives formulas for almost all combinations of boundary conditions and loads for circular plates and rings in Table 24.

10.3 RECTANGULAR PLATES

As stated in 10.1, moments at any point due to any load are in both perpendicular directions, so are stresses. For symmetric loads, the moment about the plane parallel to the larger side (M_a hereafter referred to as *M*) is higher than other M_b as shown in Figure 10.2.

10.3.1 EDGES SS

Resistance of support along the smaller side is negligible if larger side > 5 times the smaller side, and stresses are calculated as the beam of span equal to the small size. For plates within the above limits of width to length ratio, beam moments will be less due to the basics explained below.

Consider a strip of unit length of plate thickness *t*, width *b* smaller side under pressure with all edges SS.

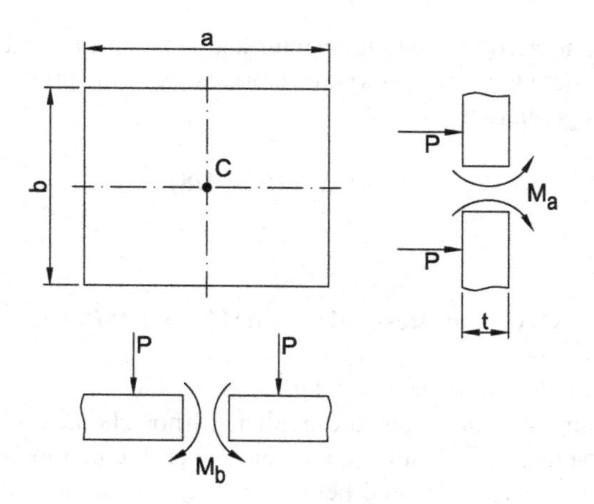

FIGURE 10.2 Rectangular plate.

If the supports at smaller edges are freed, it is a SS plate beam with span b.

Max deflection (y) and moment (M) at center point *C* as per beam formulas are

$$y = 5P\,b^4/(384EI), \quad M = P\,b^2/8, \quad \sigma = 0.75P(b/t)^2$$

Actual deflection and moment at the center of the plate are less due to the fact that the deflection of the plate from *C* along the length both sides reduces and is zero at edges due to boundary conditions. Thus, support at the other edges offers resistance and induces moment along the length which is maximum at the edge and minimum at *C*, and is proportional to n. Maximum resistance is observed for n = 1 (square) and reduces for lower values of *n*. For any value of *n*, the maximum stress is at the center.

The *equation* for maximum *M* at *C* (plane along length at a) for the rectangular plate under pressure with SS edges with v = 0 is given by Westergaard as

$$M = \left\{1/\left[8\left(1+2n^3\right)\right]\right\} Pb^2$$

Moments are induced in both directions of the plate (Ma and Mb, Ma > Mb) and affect each other for metals with v > 0, and the relationship is given below:

M with v > 0 = [(M with v = 0) + v(M in the other direction with v = 0)]

Thus Max M = $M_a - v\,M_b$

10.3.2 Fixed Edges

A plate with n < 0.5 resistance is almost negligible due to the basics explained in 10.3.1, and the maximum moment at the middle of the longer side edge is the same as that of the fixed beam and = P $b^2/12$. Stresses are higher along width compared to the length.

10.3.3 Equations for Maximum Stress for Rectangular Plates

The general equation for maximum bending stress for rectangular plates is given by Eq. 10.19

$$\sigma = M/\left(t^2/6\right) = k\,P(b/t)^2 \quad (10.19)$$

where k = constant depends on the side ratio (n = b/a) and boundary conditions.

Values of k for SS and fixed edges under pressure are given in tables 26.1 & 2 (Ref. 2).

Values of k for SS edges of plate under pressure are

0.287, 0.337, 0.395, 0.462, 0.534, 0.61, 0.662, 0.723, and 0.75 for values of n (b/a) of

1 0.9 0.8 0.7 0.6 0.5 0.4 0.3 $\leq$0.2.

The following equations can be used for calculating the *k* value in Eq. 10.19

$k = -0.5004 + 1.4328n - 1.195n^2 + 0.8156n^3 - 0.31n^4 + 0.047n^5$ for n between 0.5 and 1

$k = -0.0414 + 0.5288n - 0.1197n^2 + 0.0091n^3$ for n between 0.2 and 0.5.

Values of k for fixed edge of plates under pressure are

0.308 0.353 0.399 0.441 0.478 0.5 for values of n (b/a) of

1 0.9 0.8 0.7 0.6 0.5

The following equations can be used for calculating the *k* value in Eq. 10.19

$k = 0.3432 - 1.8604n + 3.9625n^2 - 3.059375n^3 + 1.0625n^4 - 0.140625n^5$ for the edge

$k = 0.705 - 2.83515n + 4.546875n^2 - 3.215625n^3 + 1.078125n^4 - 0.140625n^5$ for the center

Comparing Eq. 10.19 and code equations for stress: SS rectangular plate 100 × 200 × 10 mm under pressure 1 MPa

$0.61*1*(100/10)^2 = 61$ MPa, as per Eq. 10.19

$Z\,C\,P(d/T)^2 = 2.2*0.3*1*(100/10)^2 = 66$ MPa, as per code

where Z = 3.4 – 2.4d/D = 3.4 – 2.4*100/200 = 2.2

Eq. 10.19 is applicable for any shape, size, and different boundary conditions of each side. *k* depends on all the variations and is given in various references. Refer Table 26 of Ref. 2 for moments and stresses in rectangular plates with any load and any boundary conditions. However, if n is less than the limiting value, the plate can be analyzed as a beam for any load and any boundary condition. A flat rectangular plate with a rectangular concentric opening in it can be analyzed as four rectangular plates separately with applicable boundary conditions and loads as per the relevant procedure explained above.

10.4 CIRCULAR RING

Equations for moments in both radial and tangential directions are derived similar to the circular plate but more complicated due to twice the number of boundaries (inside and outside). Equations are available in Ref. 2 (Table 24, case 1 and 2) and applicable to unstayed TSs, end closures of jacketed vessels, flat flanges, and similar components.

The general equation for the maximum moment in circular rings from plate theory for any load and any boundary condition is given by Eq. 10.20.

$$M = k\,P\,a^2 = k\,W\,a = k\,Mo \qquad (10.20)$$

where

Loads P, W, and Mo are pressure, point, load and moment

a and b are the outside and inside radius of the ring

k is a constant that depends on n = b/a and boundary conditions, available in Table 24 of Ref. 2

Approximate equations for k values are formulated based on Ref. 2 and tabulated for various loads and boundary conditions in Table 10.1. Maximum stress in circular ring components used in pressure vessels is given in the following sections.

TABLE 10.1
k Equation in $M = k\,P\,a^2 = k\,W\,a = k\,Mo$ for Flat Circular Rings, W at b

Outer	Inner	Load	Equation for k, n = b/a = 0.1 to 0.9
fixed	fixed	p max M at b	$0.21867 - 0.885n + 1.39313n^2 - 0.79792n^3$ (n = 0.1 to 0.7)
fixed	noz(g)	p max M at a	$0.1247 + 0.007579n - 0.4413n^2 + 0.36208n^3 - 0.053125n^4$
fixed	free	p max M at a	$0.12098 + 0.064042n - 0.26088n^2 - 0.20417n^3 + 0.05817n^4$
ss	fixed	p max M at b	$0.38015 - 1.58016n + 2.52119n^2 - 1.44854n^3$ (n = 0.1 – 0.7)
ss	noz(g)	p max M at b	$0.3205 - 0.0955n - 1.202n^2 + 1.425n^3 - 0.44922\,n^4$
fixed[1]	free	W max M at b(-at a)	$-0.68233 + 14.13n - 57.737n^2 + 81.256n^3 - 0.8138\,n^4$
fixed	noz(g)	W max M at a	$-0.00277 + 0.56033n - 0.5488n^2 - 0.083333n^3 + 0.07552\,n^4$
ss[1]	noz(g)	W max M at b	$-0.0012375 + 6.504n - 27.1268n^2 + 3793n^3 - 17.5\,n^4$
Ss	free	W max M at b	$0.10747 + 2.69377n - 4.32323n^2 + 3.923n^3 - 1.41146\,n^4$
ss[2]	free	Mo at b, max M at b	$2\text{-}16.848n + 85.4965n^2 - 159.24n^3 + 105.966\,n^4$
ss[2]	free	Mo at a, max M at b	$3\text{-}16.848n + 85.4965n^2 - 159.24n^3 + 105.97\,n^4$

p: pressure, W: conc. Load, Mo: moment, a: outer radius, b: inner radius, g: guided, and ss: simply supported

Note 1: k values are valid for n = 0.1,0.3,0.5,0.7, and 0.9; note 2: k value is not valid for 0.1 < n < 0.3

10.4.1 Flat Ring under Internal Pressure with Both Edges Supported

Figure 10.3 shows an arrangement of jacketed vessel consisting two concentric cylinders (radius a and b, a > b) connected by two pairs of flat circular rings (unstayed tubesheets) at a finite distance and the enclosed space is under pressure. The loading diagram is identical for both. AB can be considered as a plate beam of unit circumference with A and B SS/fixed boundary. Deformation is shown in dotted lines. The expressions for both moments available at any point of the plate are given in Ref. 2 (Table 24, case 2c, d, g, and h). The effect of difference in elongation in both shells is neglected. Each plate is considered as supported at the point of its joint with both shells. The boundary condition is either SS or fixed whichever is more appropriate and depends on the stiffness of shells at the joint compared to the plate. If the shells allow edge rotations, SS is appropriate and fixed if not allowed. Shells hardly allow rotation. Hence, both fixed edges are nearer to the practical situation. However for conservative design, one of the edges may be assumed as SS. However, the case numbers in Table 24 in Refs. 2 are given below for each combination of support type:

The code gives an empirical expression as

$$t = 0.707\sqrt{P\,b(a-b)/S}$$

Assuming outer edges fixed due to a higher diameter, calculations provided for comparison with the code equation for P = 1 MPa, a = 500, b = 250, t = 20 mm, v = 0.3, case number 2g

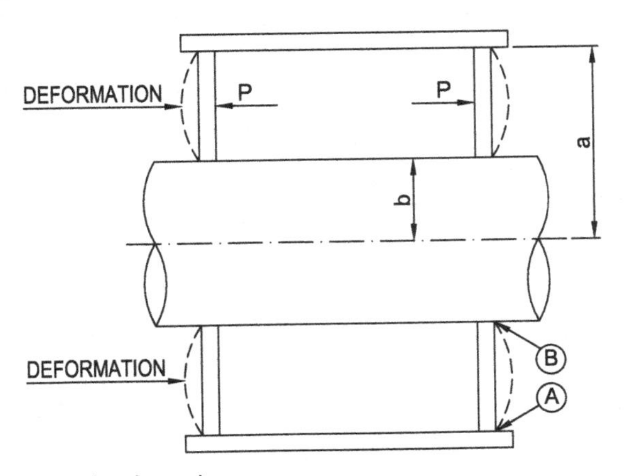

FIGURE 10.3 Flat circular ring pair.

Outside Support	inside support	case number	K_m for n = 0.5
SS	SS	2c	0.1223
Fixed	SS	2g	0.0262
SS	fixed	2d	0.0393
Fixed	fixed	2h	0.0247

Max $M_{ra} = 0.0262\ P\ a^2 = 6550$ N, stress= $M_{ra}/(t^2/6) = 98$ MPa
Stress as per code = 78 MPa

10.4.2 Flat Ring under Internal Pressure with One Edge (Outer) Supported

Figure 10.4 shows an arrangement (similar to 1. jacketed vessel, 2. shell and channel of boiler) of two concentric cylinders closed by a flat circular ring under pressure. The dotted line shows the deformation of the ring. The expressions for moments in both directions at any point of the plate are given in Ref. 2 (Table 24, case 1 and 2 b, f, i, j). In this arrangement, the plate is subjected to concentrated load (F) through the circumference of inner cylinder (inner edge of plate) apart from pressure over its surface and the boundary condition is nearer to the guide. The boundary condition of outer edge is nearer to that of the fixed edge but can be taken as SS for conservative design. The case numbers in Table 24 in Ref.2 with inner edge guided, outer edge (OS) SS or fixed (f) for loads force (F) or pressure (P) are: 1b for (OS = SS, load = F), 2b for (OS = SS, load = P), 1f for (OS = f, load = F), and 2f for (OS = f, load = P).

The code gives expression for stress by

$$t = 1.414\sqrt{P\,b(a-b)/S} \text{ or } \sigma = 1.414^2\,Pb(a-b)/t^2$$

For (P = 1 MPa, a = 500, b = 250, t = 20 mm, v = 0.3), boundary conditions: inside guided and outside SS and F.

$$F = \pi b^2 P/(2\pi b) = 125 \text{ N/mm}$$

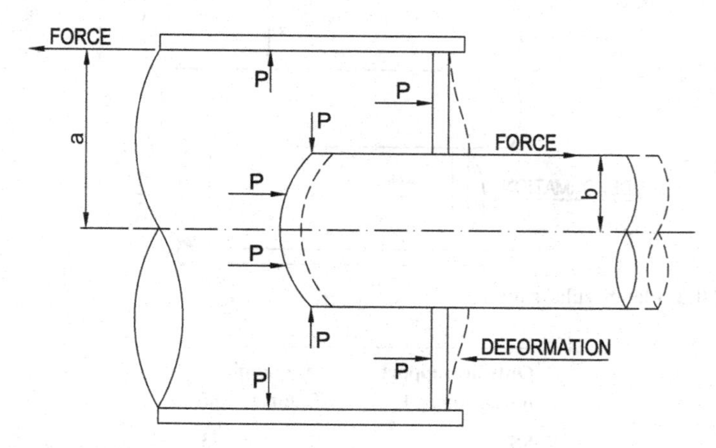

FIGURE 10.4 Flat circular ring.

1b and 2b-max M_{rb} = 0.3944F.a + 0.1223P.a^2 = 24650 + 30575 = 55225 N, stress = 6 M/t^2 =828

1f and 2f-max M_{ra} = 0.1345F.a + 0.0601 P.a^2 = 8406 + 15025 = 23431

1f and 2f-max M_{rb} = 0.2121F.a + 0.0407 P.a^2 =13256 + 10175 = 23431, stress 6M/t^2 =351 MPa

Stress as per code is 312 MPa

For design programs, equations for approximate k values in Eq. 10.19 are given in Table 10.1.

10.5 STIFFENERS

As the size of flat plate increases, the thickness increases for the same load and boundary conditions at a much higher rate and proportion to the square of the size. In rectangular vessels, opposite plates can be stayed by stay bars.

For higher sizes, stiffeners are used to reduce the thickness and weight. Stiffeners are arranged in such a way so that the stress in unsupported portion of the plate (area bounded by stiffeners) as per Eq. 10.19 is within allowed limits.

Stiffeners are of three types.

1. Main stiffeners with ends connected to the support (edges of plate) which will take load from second type stiffeners as well as directly from the plate.
2. Internal stiffeners which will take load from the plate and transfer to main stiffeners.
3. A square pattern of small size is generally used for very low pressures to increase resistance to vibration (to increase natural frequency).

Each stiffener loading diagram and boundary condition can be arrived by fundamentals and logics. When the plate size is much larger, both types of stiffeners are used, and the optimum arrangement requires several trials due to various factors involved. The designer requires practical expertise together with basics to avoid trials to obtain an economic design. Stiffeners are full welded to have uniform temperature distribution when the plate is under higher temperature. For ambient and low temperatures, staggered welding can be adopted. Stiffener ends are welded to a connecting stiffener or any member with sufficient weld area to avoid shearing of its weld with plate at the ends. The weld design is explained below

For the purpose of stiffener design, a flat plate is considered as supported at the edges (E) by the plate corner welded to it and generally perpendicular to it as shown in Figure 10.5. The edges and stiffeners are considered as structural frames and plate portions enclosed by frames are panels. Unless stiffener ends are welded to the connecting stiffener, the above consideration is not true. When welded thermal stresses are induced in case the temperature difference along the depth of the

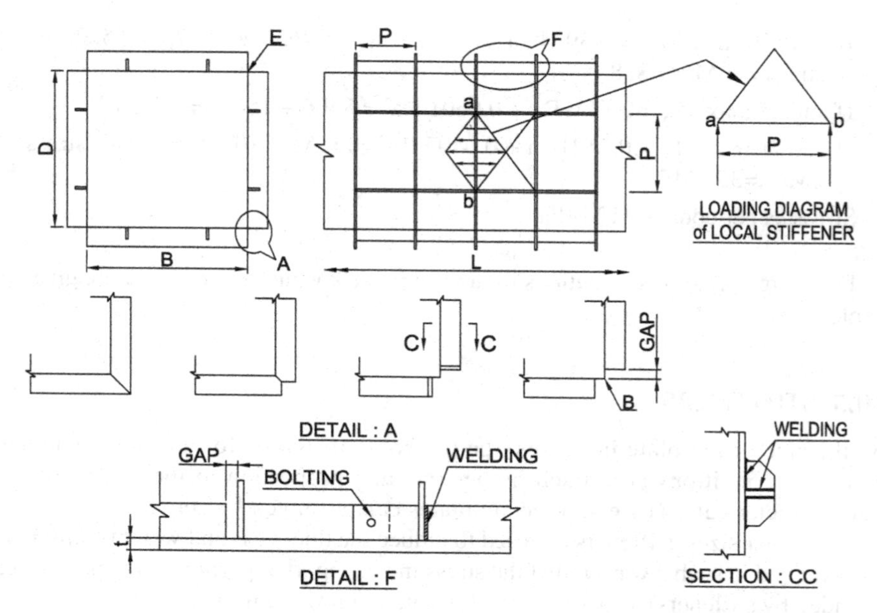

FIGURE 10.5 Stiffeners on the rectangular vessel.

stiffener exceeds a certain limit. To avoid thermal stresses, a small gap is left due to which pure shear stress is induced in the plate as shown at edge (B) in Figure 10.5. If the gap is more than its thickness, bending stress is also induced. The shear stress is equal to the reaction of the beam of the combined section of the stiffener with the effective width of the plate. For normal span (length of stiffener), the shear stress will be within the limits, but the welding between the stiffener and the plate is subjected to horizontal shear stress as shown in Figure 3.8 and described in 3.7.3.2. It is similar to a fabricated beam. The Shear force is maximum at ends and shearing may start from the end if the weld is inadequate. The design philosophy to overcome this factor is to weld stiffener to stiffener or weld through the additional end plate to get the required weld area at the stiffener end as shown in Figure 10.5. To reduce thermal stresses in stiffeners, the welding between stiffeners shall be limited to required depth and shall be within insulation as shown in Figure 10.5a. Rule 21 of EN13445-3:2009 gives the equation for the required weld area for a similar arrangement.

Code provide an analytical procedure to calculate stresses for only a limited choice of stiffeners. However, there is no limit of arrangements.

10.6 RECTANGULAR VESSELS

The design of a rectangular vessel shown in Figure 10.5 is illustrated with example 10.1

Example 10.1: Design stiffeners for a rectangular vessel of size B × D × L = 1.5 × 1.5 × 10 m, thickness t = 3 mm, pressure P = 0.02 MPa, and basic allowable stress s = 120 MPa.

Corners are considered as SS.

Membrane stress due to P, σ_m = P (b or d)/(2t) = 5 MPa

Bending stress in each of four side plates as per Eq. 10.19 = 0. 5*0.02(1500/3)^2 = 2500 MPa considering fixed boundary. As per 10.3.3, k = 0.5 (n = B/L or D/L = 1.5/10 < 0.5)

Because allowed stress for membrane + bending stress is 1.5S, bending stress is to be limited to 1.5*120–5 = 175 MPa. Therefore as per code or above theory, the plate cannot withstand pressure. Stiffeners are to be provided at a pitch of about 300 mm or stay bars are to be provided inside. If stay bars may not be used due to temperature and fluid, also 300 pitch is too close and codes do not have alternate solution. Hence, stiffeners are designed from basics described in 10.5. Using Eq. 10.19, the approximate size of the plate to withstand pressure is 500 × 500 mm. Therefore, a main stiffener of 100 × 6 × 1500 at 500 pitch along the length and 2nos. of 50 × 3 × 500 between each main stiffener is selected as shown in Figure 10.5

Calculation of moment and stress in stiffeners by structural basics with the above data.

1. Local stiffener: as per loading diagram shown in figure

   M = 0.02*500^2*500/6 = 416667 Nmm

   Section modulus required Zr = M/(1.5S) = 416667/180 = 2315 mm^3

   To calculate available combined Za, plate effective width is approx

   40t or = $\sqrt{t}$(1.11E/Y) = $\sqrt{3}$(1.11*192000/200) = 103 mm as per code.

   Za = 10092 plate side and 2408 mm^3 on the stiffener side

   Bending stress in stiffener = σ_b = M/Z = 416667/2408 = 173 < 180 hence safe.

   Bending stress in plate = σ_b = M/Z = 416667/10092 = 42 MPa

   Membrane + bending stress = $\sigma_m + \sigma_b$ = 5 + 42 = 47 < 180 hence safe.

2. Main stiffener considering edges SS M = PB^{2}p/8 = 0.02*1500^2*500/8 = 2812500 Nmm

   Za = 29362 plate side and 15566 mm^3 on the stiffener side

   Bending stress in stiffener = σ_b = M/Z = 2812500/15567 = 180 < 180 hence safe.

   Bending stress in plate = σ_b = M/Z = 2812500/29362 = 96

   Membrane + bending stress = $\sigma_m + \sigma_b$ = 5 + 96 = 101 < 180 hence safe.

Stiffener design as per code with only the main stiffener is almost the same as the theory in this section.

10.7 TUBE SHEET

10.7.1 General

TS is a perforated flat plate attached to a shell at the outside and to tubes at perforations inside the shell as shown in Figure 10.6a.

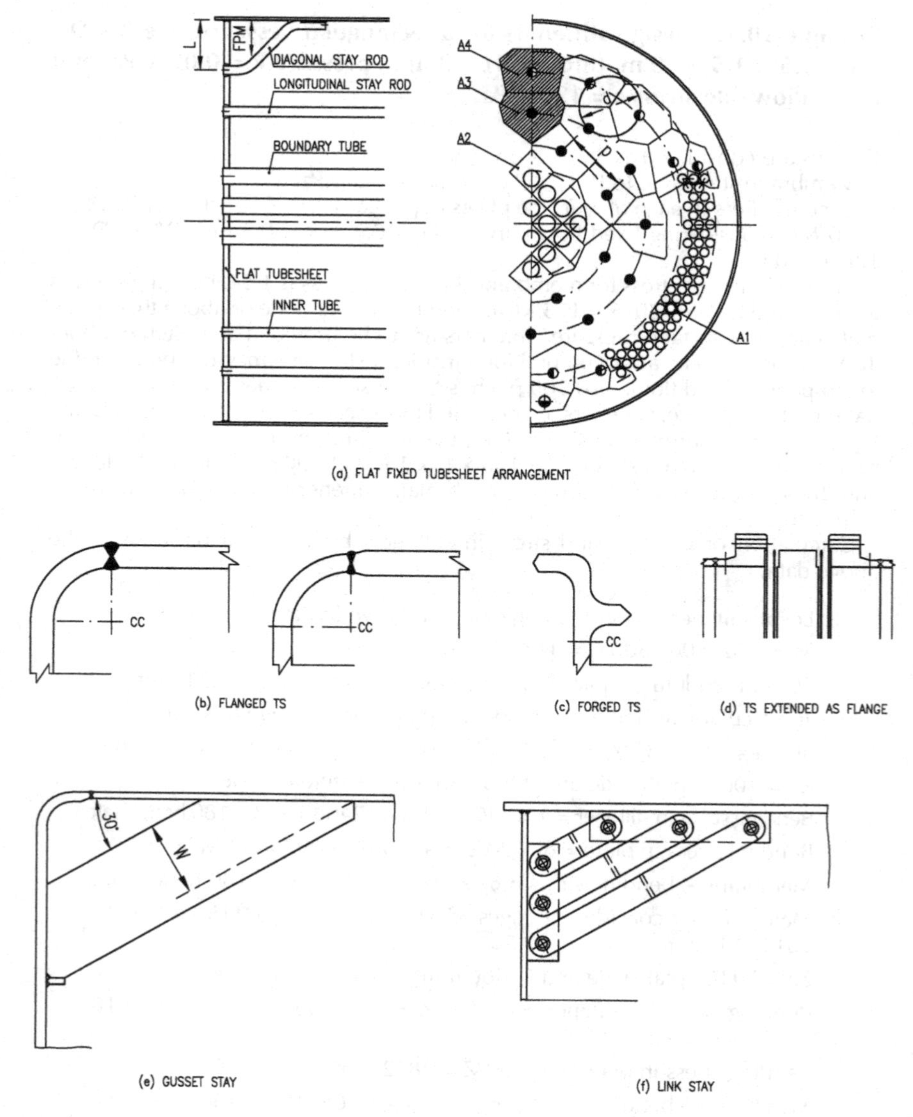

FIGURE 10.6 TS and its staying arrangements.

Normally TSs are of three types.

1. Both sides under pressure, stayed by tubes and the other arrangement against pressure as in the shell and tube type fixed-TS exchanger or boiler.
2. Both sides under pressure as in stationary TS in a shell and tube type floating TS or U-tube exchanger or fixed TS exchanger or boiler with tubes attached to TS only for sealing (unstayed).

3. Both sides with same pressure for supporting tubes as in floating TS or intermediate TSs in any type of heat exchanger

The second type of TS is a flat plate under pressure neglecting the resistance offered by tubes and can be analyzed as shell end closure described in Chapter 5 and section 10.2. The third type is only the structural part. Analysis for the first type of TS is given in this section.

10.7.2 Attachment to Shells, Tubes, and Stays

A TS is attached to a shell by four methods

a. Directly welded, called flat TS [Figure 10.6a]
b. Flanged and butt welded, called flanged TS [Figure 10.6b]
c. Forged and butt welded, called forged TS [Figure 10.6c]
d. Extended and bolted to the shell or channel flange and, called TS extended as flange or gasketed between shell and channel flanges [Figure 10.6d]

A TS is attached to tubes in two functional ways

1. Sealing attachment: Assumed as no contribution from tubes to TS in resisting pressure: Expanded and/or seal welded and called plain tubes.
2. Strength attachment: Stays the TSs (fixed TS) from bending due to pressure by expanding with grooves and/or with belling or strength welded with or without expanding and called stay tubes.

Tube layouts in a TS are

- TS fully (about 80%) covered by tubes mostly concentrically
- Partially covered by tubes and the rest of the area stayed by longitudinal stay rods/pipe and/or diagonal stay rods or gussets as shown in Figures 10.6a and 10.6e.

Longitudinal rods, diagonal rods, or flat plate gussets are used by strength welding for staying wide spaces uncovered by tubes between the outside tube limit and the shell inside. When the gusset stay width along the TS is large, bolted link stays are used to avoid nonuniform distribution of load in gusset stays as shown in [Figure 10.6f].

10.7.3 Analysis

Flat TS analysis can be done using flat plate theory, but is complicated due to boundary conditions, perforations, varying resistance of staying arrangement, and self-limiting plastic theory. Only finite element analysis will give accurate results. With several assumptions listed below, analysis can be done by the procedure explained in 10.2 and 10.4.

For fixed TS analysis with staying, the following assumptions are made:

1. Tubes (usually regularly pitched) and their staying boundaries are isolated as it is only tube in the center of its boundary in TS. Other stays are also similarly assumed.
2. Shell contribution is only limited to the boundary of nearest staying points.
3. TS-shell joint is SS or fixed for the bending of TS. When considered as fixed, the radial distance from the shell inside to a limit depends on the thickness of TS and pressure and will not require any staying and called flat plate margin (FPM ref. IBR).
4. Tubes, stays, and shell are elongated equally due to pressure for the purpose of stress in TS. Tubes actually act as springs with a constant equal to EA/L, A = area of cross section, L = length of tube or TS to TS distance.
5. Tubes and stays are subjected to only tension (tube at its joint with TS is actually subjected to local bending).
6. The TS is thin enough that isolation of pressure areas of each stay and TS unsupported portions is almost true.

10.7.3.1 Pressure Areas of Stays and Stress in Stays

The total pressure area of TS (leaving FPM if considered) is divided into parts equal to the number of stays. The guideline for dividing is that any point on the boundary shall be equidistance from the nearest two staying points if not more. For this purpose, a FPM circle if considered or a radial point on the TS edge or its commencement of curvature (CC) in the case of flanged or forged TS is the support point on the outer boundary. An extended flange-type gasket reaction circle (G) is considered as the support point (refer Chapter 13).

$FPM = C\, t/\sqrt{P}$, where t is the TS thickness, P is pressure, and C is constant; C = 95 exposed to radiation else102 if t is in mm, P in MPa (111, 118.4 if t is in inch and P in psi)

Stay tubes and stays are designed for tension due to pressure on area in its boundary as drawn with the above guidelines deducting the area covered by stay as shown in Figure 10.6a. Pressure areas as shown in the figure are:

A_1 = for stay tube within tube nest

A_2 = for boundary stay tube

A_3 = for longitudinal stay bar

A_4 = for diagonal stay bar or gusset

Calculation of tensile stress in staying parts is simple and equal to the pressure force divided by cross section of staying part and same for stay tubes and longitudinal stay bars. Pressure force $F = P\,(A_1, A_2, A_3)$ as given in example 10.2. Pressure force in diagonal or gusset or link stays is different due to the angle (θ) between shell wall and longitudinal axis of these stays and given by

$$F = P\, A_4/\cos\theta$$

The cross-sectional area of the gusset is the effective normal section as calculated in example 10.3. The optimum angle is 30°.

Example 10.2: P = 1 MPa, area of bar = 1000 mm², gross pressure area including stay A_3 = 101000 mm², calculate tensile stress in the longitudinal stay bar.

Net pressure area for bar = 101000–1000 = 100000

Tensile stress in bar = 1*100000/1000 = 100 MPa

If tube/pipe with 100 mm OD and 3mm thickness is used in the place of the bar

Tube area $\pi d^2/4$ = 7850 mm²

Cross-sectional area = appx. 3.14d.t = 942 mm²

Tensile stress in tube = 1(101000–7850)/942 = 98.8 MPa.

Tensile stress in stay tubes is calculated in the same way as for the longitudinal stay bar in example 10.2.

10.7.3.2 Unsupported Portions in the Tube Sheet

Several portions of TS bound by three or more staying points including the support point of the shell or CC (circles or pitch or rectangles as the case may be) are isolated, and bending stress is calculated as a isolated flat plate with fixed edges as per 10.2 or 10.3. For compiling the diameter *d* of each such portion, draw a circle (diameter *d*) touching three or more staying points provided all points are not in one half of the circle. The other method is maximum pitch (p) between any two staying points in any direction (approximately same as *d*). Pitch at the boundary for the purpose = L/1.5 considering TS to the shell joint as fixed, where L = actual distance from the boundary staying point to the shell as shown in figure. To derive moment and stress in each portion, consider *d* or *p* as the span of a fixed plate beam in which conservative moment and stresses are given by

$$\sigma = M/Z = \left(P\ p^2/12\right)/\left(t^2/6\right) = P(p/t)^2/2 = P(p/t)^2/C$$

In codes, the same equation is used with approximately the same value of C(2).

IBR code: *d* is used in place of *p*, C = 2.5 to 3.2 depending on the degree of fixity of staying point. *d* may be = $\sqrt{(A^2 + B^2)}$ where A and B are rectangular pitches.

ASME codes: C = 2.1 for $t \leq 11$, and C = 2.2 for t > 11mm. For tubes within the tube nest where *p* is computed as = $\sqrt{p^2 - a}$, a = cross-sectional area of the staying part, because the area of the staying part is considerable, and its effect on the pressure area cannot be neglected.

Unsupported space called *breathing space* is provided between differentially expanding parts on tube sheet of boiler as per Indian and some other country codes for flexibility. Regulation 589 of IBR (Indian code) specify the minimum space as: 40 mm between tube and shell, 100 tube and stays, 50-100 (5% of shell ID) tube/shell and furnace tube, and 150-250 stays and furnace depending on length of furnace.

10.7.3.3 Diagonal Stay and Gussets

Longitudinal stays are not advisable where inside maintenance is required. Furthermore, when the tube length is large, longitudinal stays are costly. An alternate is diagonal stays. Diagonal stays can be bar or plate. They are connected to the shell at an angle of about 30°. A gusset stay as shown in Figure 10.4e is preferred when more diagonal stays are required in one radial line. If the gusset width is too large, split it into 2 or 3 and provide a link type as shown in Figure 10.6f. The pressure area for the gusset is the same as the diagonal stay. Diagonal stay is subjected to local bending stress if the diagonal line extended joining point in TS is outside the diagonal bar.

Example 10.3: Instead of the longitudinal stay bar, 30°-diagonal stay bar of the same area 1000 mm^2 is used, and tensile stress is calculated in the diagonal bar

Tensile force = 1*100000/cos30 = 116628 N and

Tensile stress = 116628/1000 = 116.28 MPa

For 30° gusset stay, the procedure is the same as that for the diagonal bar. The cross-section area applicable is not at the TS joint or at the shell joint, but minimum across along the diagonal length (w × t) as shown in Figure 10.6.

10.7.3.4 Flanged and Extended as Flange Tube Sheets

Flanged knuckle portion of flanged TS will transfer the force on it to the shell and will withstand pressure. If the thickness of TS is not less than the minimum required for the shell, it is safe. Else, finite element analysis may be required as there is no reasonably accurate analytical procedure available.

Stress in the flanged portion of stayed TS as shown in Figure 10.6d is due to the flange moment $W\ hg$ (refer Chapter 13) and is calculated by Eq. 13.2. [1.91 $M_d/(G\,t^2)$]

Unstayed TS extended as a flange is the same as a blind flange and the stress induced is the sum of the stresses due to the flat circular plate with a gasket reaction diameter as the diameter (refer section 10.2) and due to flange moment $W\ hg$ (refer Chapter 13) which is calculated by Eq. 13.3.

$$\left[1.91M_a/\left(Gt^2\right)+0.33P(G/t)^2\right]$$

Code in addition to the above simple analysis gives detailed analysis of flat TS with a particular tube layout and without stays are given but result in a higher thickness.

REFERENCES

1. Code ASME S I & VIII D-1, 2019.
2. Roark, R. J. and Young, W. C. *Formulas for stress and strain*, 5th Edition.

11 Supports

Supports will induce local stresses in pressure vessels like any attachment. Attachments transfer external loads, while supports will take the loads from the shell and transfer to the foundation or structure. Local stresses are induced due to reaction from the support. Attachments are generally of small size compared to the shell.

Calculations of stresses in some of the normal support components are by simple structural analysis. The analysis of certain support components such as saddles for horizontal vessels and skirts for vertical vessels is not simple and described in this chapter.

11.1 SADDLE

Horizontal cylindrical shells are supported by two or more supports called *saddles*, one fixed rest sliding. The saddle as shown in Figure 11.1 normally covers 120° (saddle angle) circumference of the shell, and 150–300 mm longitudinal length of the shell. Apart from pressure and gravity loads, the structural effect of wind/seismic is to be applied. The analysis consists of the following:

- Longitudinal membrane plus bending stresses in the shell at the saddle section due to pressure, gravity, and wind/seismic loads.
- Transverse shear stresses at the saddle section due to gravity and wind/seismic loads.
- Circumferential stresses at the saddle section due to the saddle reaction.
- Design of saddle parts.

The saddle being rigid compared to the shell later deforms locally due to reaction force for gravity plus wind/seismic loads, and local stresses are induced around the saddle

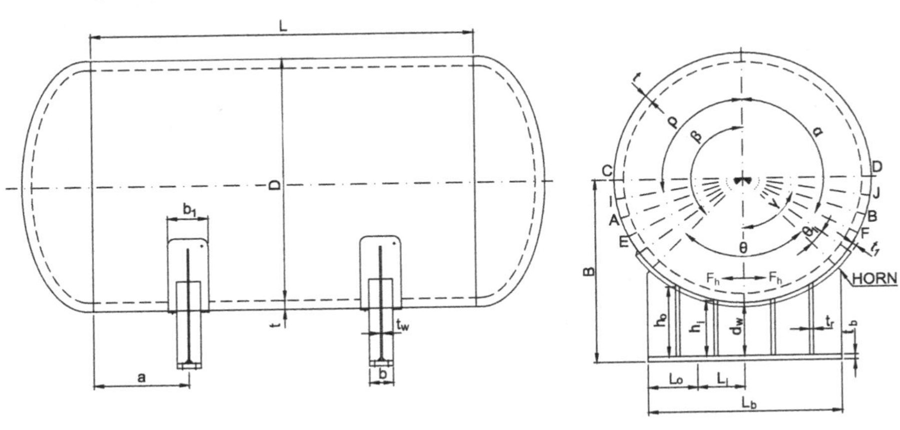

FIGURE 11.1 Shell with saddles.

DOI: 10.1201/9781003091806-11

periphery. The local stresses are calculated based on linear elastic mechanics and cover modes of failure by excessive deformation and elastic instability. The analysis is not simple and carried out by the integration method. The analysis is called Zick analysis.

The following are different arrangements influencing the calculation of stresses at the saddle section:

1. Simple saddle
2. With stiffeners in plane of the saddle inside or outside (Figure 11.2a–c)
3. With stiffeners both sides of the saddle inside or outside (Figure 11.2d and 11.2e)
4. Shell stiffened by head (torispherical, elliptical, and flat) if $a \leq R_m/2$

For the purpose of stress calculations:

Stiffeners are considered as in plane, if $h \leq y$

Stiffeners are considered as both sided of saddle, if $y < h \leq Rm$

Stiffeners are not effective for the saddle section, if $h > Rm$

where $y = 1.56\sqrt{(R_m t)}$

h = distance of the saddle section to stiffener ring

Rm = mean radius of the shell

t = thickness of the shell.

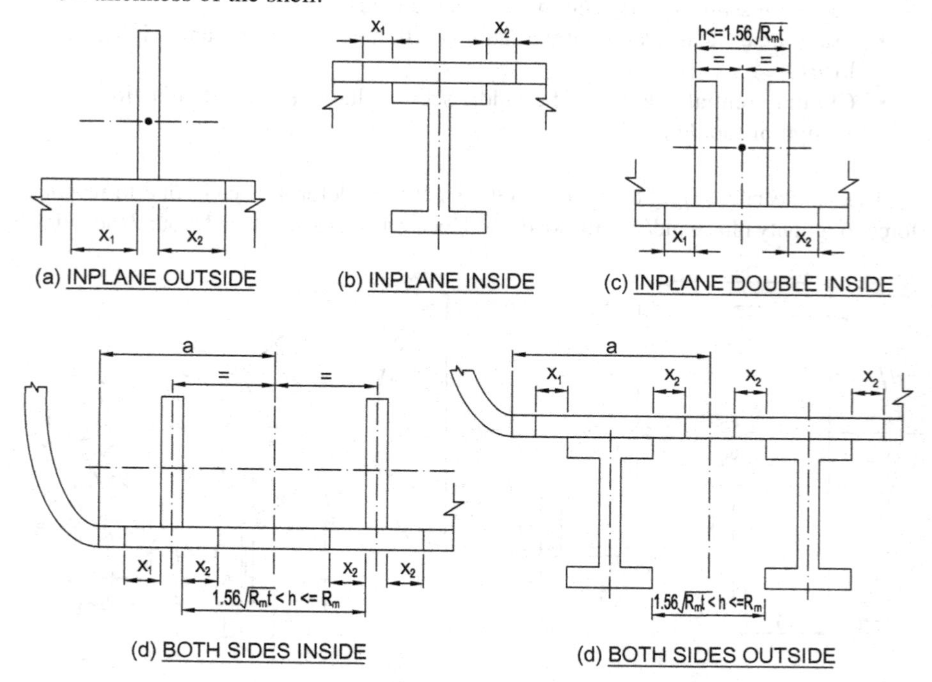

FIGURE 11.2 Stiffeners on the shell at the saddle section.

11.1.1 Longitudinal and Transverse Shear Stresses in the Shell

The following steps explain the analysis. Refer Figure 11.1 and example 11.1 for notation.

Step 1: Calculate reaction Q, shear force T, and bending moment M_1 at each saddle section due to gravity and wind/seismic loads by beam theory. If the support distance is abnormally large, moment M_2 in between supports may be more than at supports and to be calculated. Q, T, M_1, and M_2 for gravity loads are computed by simple beam theory with the loading diagram for two saddles. For three saddles due to an abnormally long shell, compute by theorem of three moments (Alapeyron's theorem). If saddles are more than three saddles, loads are distributed among saddles appropriately from the loading diagram, or software such as STAAD is used for computing Q, T, M_1, and M_2. For two-saddle symmetrical shells with a uniformly distributed load, the following equations are computed considering as overhang beam-using-beam formulas.

$$Q = W/2$$
$$M_1 = [-W a'^2/L']$$
$$M_2 = W[(L'/2 - a') - L'/4]$$
$$T = [W \max(2a'/L', 1 - 2a'/L')]$$

where

$L' = L + 2L_1$

L_1 = equivalent length of end closure for the purpose, flat = 0, dished = 2H/3

H = height of the dish

a = distance from the saddle section to dish (other than hemispherical) or flat head

W = total load

M_1 = moment at the saddle section

M_2 = max moment between saddles

T = shear force

$a' = a + L_1$

The structural effect of wind/seismic is limited to Q only as the shear and moment are in the horizontal direction and generally neglected compared to gravity loads. For the calculation of reaction due to wind/seismic, refer example 12.6, and transverse force is described in 11.1.3

Step 2: Calculate bending stress (M/Z) due to M_1 and M_2 and add longitudinal pressure stress to obtain membrane + bending stress and tangential shear stress T/A due to T.

where

Z = section modulus = $K \pi R_m^2 t$ for the thin cylinder

A = shear resistance = $K \pi R_m t$ for the cylinder

K = constant for the local effect (reduction in resistance) of the saddle causing distortion in circularity and depends on saddle angle θ and type of stress.

Except the portion covered by saddle angle θ plus (30°–θ/12) either side, the circumferential section at the saddle is ineffective and does not contribute in resisting moment of inertia.

When the shell is stiffened by providing stiffener rings or stiffened by end closures other than the hemispherical head, the shell circularity is considered as unaffected by the saddle reaction.

Similarly, transverse shear is unaffected with the stiffener in plane. Stiffeners on both sides of the saddle or stiffening heads will not resist shear Force *T*. However, shear stress is induced due to saddle reaction Q stiffened by heads and without stiffener ring.

K is derived by integration (Ref. 2), and its values for bending stress (σ_3) at top (K_1), for (σ_4) at bottom (K_1'), and shear stress τ (K_2) are obtained. The equations are given below.

$$K_1 = [\Delta + \sin(\Delta)\cos(\Delta) - 2\sin^2(\Delta)/\Delta]/[\{\sin(\Delta)/\Delta - \cos(\Delta)\}\pi] \quad (11.1)$$

$$K_1' = [\Delta + \sin(\Delta)\cos(\Delta) - 2\sin^2(\Delta)/\Delta]/[\pi\{1 - \sin(\Delta)/\Delta] \quad (11.2)$$

$$K_2 = (\pi - \alpha + \sin\alpha\cos\alpha)/(\pi\sin\alpha) \quad (11.3)$$

The shear stress (τ) is due to saddle reaction Q when stiffened by heads = $Q/(K_3\, \pi R_m\, t)$; K_3 is given by

$$K_3 = (\pi - \alpha + \sin\alpha\cos\alpha)/[\sin\alpha(\alpha - \sin\alpha\cos\alpha)] \quad (11.4)$$

where $\Delta = \pi/6 + (5/12)\theta$, $\alpha = 0.95\beta$, and $\beta = \pi - \theta/2$

Also longitudinal membrane and shear stresses are induced in stiffening heads. Constant K_3 is applicable for shear stress and longitudinal membrane stress (σ_5) due to $Q = Q\, K_4/R_m\, t_h$, the K_4 factor is similarly derived and given by Eq. 11.5, $R_m\, t_h$ = resisting area, and t_h = head thickness.

$$K_4 = 3/8(\sin\alpha)^2/(\pi - \alpha + \sin\alpha\cos\alpha) \quad (11.5)$$

Example 11.1 gives the detailed analysis.

11.1.2 Analysis of the Saddle Reaction due to Wind/Seismic

Because both wind and seismic cannot occur, the max reaction of wind or seismic is added to reaction due to gravity load. The analysis is simple as covered in Chapter 12, except the reaction due to transverse force. Transverse force acting at CG will induce moment (M) at the saddle base and equal to force multiplied by the height of CG from the base. The reaction due to this moment is not uniform along the length of the saddle. It is max at the outer edge and zero at middle. The max reaction at the edge is derived in 11.5.1. For the purpose of calculating Q for saddle analysis, max reaction can be calculated by taking moment per unit width at the middle of saddle and equated to M as shown in Figure 11.3

$$M = 2\left[\left(\frac{1}{2}p\frac{L}{2}\right)\left(\frac{2}{3}\frac{L}{2}\right)\right] = p\frac{L^2}{6} \quad (11.6)$$

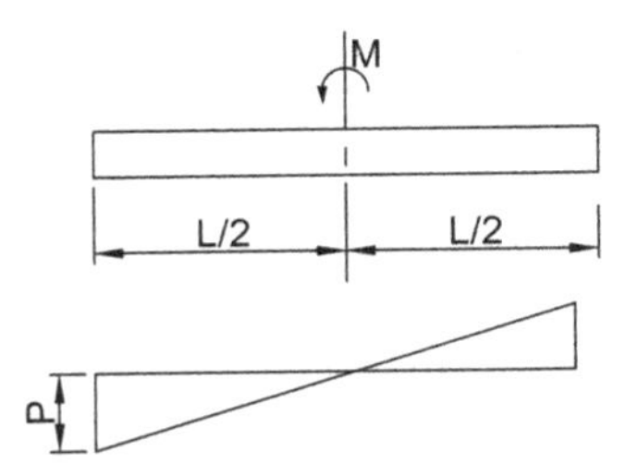

FIGURE 11.3 Saddle reaction due to moment.

where p = pressure per width B and unit length and Q = p L/2, the above equation is reduced to Q = 3M/L. That is, the maximum reaction is 3 times the average. The total reaction is sum due to gravity and wind/seismic loads. Calculation of Q due to wind/seismic for the shell with two saddles is illustrated in Example 12.6.

11.1.3 Circumferential Stresses at the Saddle Section due to Saddle Reaction Q

The total reaction due to gravity and wind/seismic loads induces circumferential membrane stress (σ_6) and membrane plus bending stress in shell at the saddle section, max at horn of saddle (σ_7), or end of wear plate (σ_{71}). The moments are derived by the integration method and proportional to $Q\ R_m$, and membrane and bending stresses are computed and proportional to Q/resisting area and $Q\ R_m/Z$ (Z for unit width). Constants are functions of angles α for membrane, β (180 – θ/2) for stress at horn and β_1 (180–θ/2–θ_1) for stress at the end of the wear plate. When stiffener rings are provided, stresses are generally reduced in the shell. Membrane stress and bending moment are the same in the shell and ring. However, bending moment induced is opposite stress inside and outside. The equations for stresses are proportional to Q/area and $Q\ R_m/Z$, and constants are derived by integration (Ref. 2) and are a function of angle β and ρ. Providing a stiffener ring inside is an advantage as the shell is in compression. However, stress in the ring will be high for bar stiffeners and can be reduced by providing an inverted T-section to obtain combined Z in tension and compression almost equal. The detailed calculations are illustrated in example 11.1 for the unstiffened saddle and 11.2 for the stiffened saddle. Equations for constants are given below (Refs. 1 and 2).

$$K_5 = (1+\cos\alpha)/(\pi-\alpha+\sin\alpha\cos\alpha) \text{ for } \sigma_6 \quad (11.7)$$

$$K_6 = [3/4Bc(Bs/B)^2 - 5/4Bs\,Bc^2/B + Bc^3/2 - Bs/(4B) + Bc/4 - B$$
$$Bs\{(Bs/B)^2 - 0.5 - \sin(2B)/(4B)\}]/[2\pi\{(Bs/B)^2 - 0.5 - \sin(2B)/(4B)\}], \quad (11.8)$$
$$Bs = \sin\beta,\ Bc = \cos\beta,\ B = \beta \text{ for } \sigma_7$$

Equations for other constants are given in examples 11.1 and 11.2

Example 11.1: Calculate local stresses in the cylindrical vessel due to the support saddle (Figure 11.1)

Design data: (units N, mm UOS) design pressure P = 1 MPa, temperature = 250°, material for shell and wear plate SA-516 70 with allowed stress at temp = 138

D = shell inside diameter = 2300

t = shell thickness = 12

t_1 = wear plate thickness = 12

b = saddle width = 250

L = 10 m length of the shell [tan to tan if dish end (hemispherical + D/6) or flat or other end]

a = 1.5 m distance from the axis of saddle to end as defined for L, if $a \le R_m/2$ dished/flat head stiffens the shell

θ = 120° circ. angle of the contact surface of the saddle support with the shell (normally 120°)

θ_1 = 5° wear plate extension angle above the horn of saddle (min. $\theta/24$ for effectiveness of t_1)

b_1 = 450 width of wear plate $[\min\{b+1.56\sqrt{R_m t}, 2a\}]$ for effectiveness of t_1

w = 800 KN half of load of the vessel without the wind/seismic effect

Q = 900 KN max saddle reaction including the effect of wind or seismic

M_1 = 180 KNm moment at the saddle support (–)

M_2 = 2800 KNm max moment between saddles

T = 560 KN shear force at saddle

Derived data required in analysis

$\beta = 180 - \theta/2 = 120°$ angle from the saddle horn to top

$\Delta = \pi/6 + (5/12)\,\theta = 80°$

$\alpha = 0.95\beta = 114°$ angle from top to max shear point just above horn

$R_m = D/2 + t = 1156$ mean radius of the shell

$Z = \pi R_m^2 t = 0.05\ m^3$ section modulus of the shell

Long stress in shell due to P, M_1, M_2: allowed= S, if stress is negative allowed stress $S_c = t\,E/(16R_m)$

$\sigma_1 = P\,R_m/(2t) - M_2/Z = -7.412$ stress between supports at top

$\sigma_2 = P\,R_m/(2t) + M_2/Z = 103.7$ stress between supports at bottom

K_1 = Eq. 11.1 = 0.107 for the unstiffened saddle (1 for stiffened)

K_1' = Eq. 11.2 = 0.192 for un stiffened (1 for stiffened)

$\sigma_3 = P\,R_m/(2t) - M_1/(K_1 Z) = 81.68$ at A and B of support (top for stiffened)

$\sigma_4 = P\,R_m/(2t) + M_1/(K_1' Z) = 29.6$ at bottom of support

Shear stress in shell due to T and Q, head (all = 0.8S) and membrane stress in stiffening head (all = 1.25S)

K_2 = Eq. 11.3 = 0.272 without the stiffener in plane (1 for the stiffener in plane)

$\tau = T/(K_2 \pi R_m t) = 47.26$ shear stress at C and D for the stiffener in plane, else at E and F

K_3 = Eq. 11.4 = 0.362 factor if stiffened by head

$\tau = Q/(k_3 \pi R_m t)$ in the shell w/o stiffener and stiffened by head at E and F angle α from top, else 0 [NA if (g = a/Rm = 1.3) > 0.5]

$\tau' = Q/(K_3 \pi R_m t_h)$ in stiffening head at angle α from top t_h = head thickness) NA

K_4 = Eq. 11.5 = $3/8(\sin\alpha)^2/(\pi - \alpha + \sin\alpha\cos\alpha) = 0.401$

σ_5 = membrane stress in stiffening head = 0 for flat head, $K_4Q/(R_m t_h) + P R_i/(2t_h C)$ for other heads, C = 1 for torispherical, h/R_i for the elliptical head, h = depth of the elliptical head, and t_h = head thickness

Circumferential compressive stress in shell without stiffener ring due to Q (membrane-allowed = S; mem + bending-all =1 .25S)

$x_2 = 0.78\sqrt{(R_m t)} = 91.87$ effective width of the shell from the saddle towards the center for resisting force and moment

$x_1 = \min(a - b/2, x_2) = 91.87$ effective width of the shell from the saddle to head for resisting force and moment

$w = \min\{b + 1.56\sqrt{(R_m t)}, 2a\} = 433.7$ min width of wear plate to resist σ_6

K_5 = Eq. 11.7 = $(1+\cos\alpha)/(\pi - \alpha + \sin\alpha\cos\alpha) = 0.762$

$\sigma_6 = -K_5 Q k/[b_2(t + f e t_r)]$ membrane stress at the base of the saddle

f = stress ratio of shell and wear plate = 1 as the same material

e = effectiveness of the wear plate = 1 (e = 1 if $b_1 > w$ and $\theta_1 > \theta/24$, else e = 0)

$b_2 = b_1 = 450$ (= b_1 if e = 1, else $b_2 = b + x1 + x2$)

$\sigma_6 = -6.348$

K_6 = Eq. 11.8 = 0.053

K_7 = If($a/R_m \geq 1$, K6), if($a/R_m \leq 0.5$, $K_6/4$) else $K_6(1.5g - 0.5) = 0.053$

$\sigma_7 = -Q/[4b_2(t + f e t_1)] - 12K7\ Q/[(t + f e t_1)^2 \min(8, L/Rm)] = -144.7$ membrane + bending stress at horn of the saddle

σ_{71} = memb + bend at tip of wear plate $[-Q/[4t(b + x_1 + x_2)] - 12K_{71}Q/[t^2 \min(8, L/R_m)]$ if e = 1 & $t_1 > 2t$, else not applicable] = not applicable as $t_1 \leq 2t$

$K_{71} = K_{61}$ If $a/R_m \geq 1$, = $K_{61}/4$ if $a/R_m \leq 0.5$, else = $K_{61}(1.5a/R_m - 0.5)$

K_{61} = equation of K_6 with β_1 in place of β

$\beta_1 = 180 - (\theta/2 + \theta_1) = 115°$

Example 11.2: Calculate local stresses in the cylindrical vessel due to the support saddle with a stiffener ring. Figure 11.2, units: N, mm, MPa

The data are the same as example 11.1: $t = 12$, $a = 1500$, $b = 250$, $\theta = 120$, $Q = 900000$, $R_m = 1156$, $\beta = 120$, and $K_5 = 0.762$

Stiffener data: Type: in-plane, inside, n = no of rings = 2

$h = 150$ = distance o/s to o/s of two rings, 0 for 1-ring, for in plane $h < 1.56\sqrt{(R_m\ t)}$, max. = R_m

$w = 150$ = contact width of the stiffener with the shell, ($w = h$ if $n = 2$ and in plane)

$w_s = h + x_1 + x_2 = 334$ = eff. width of the shell for A and Z = [if(n = 1, $w + x_1 + x_2$), if(n = 2 & in plane, $h + x_1 + x_2$), if(both sides, $w + 2x_2$)]

A_S = area of each stiffener, (inverted T-section: web = 150×10, flange = 100×16) = 3100

A = combined area of the stiffener and shell = 14210

Z_V = section modulus shell side = 1.16E6

Z_S = section modulus stiffener side = 593734

$\rho = -158.58 + 7.8668\theta - 0.088037\theta^2 + 4.3011E-4\ \theta^3 - 8.0644E-7\ \theta^4 = 93.71$

Circumferential compressive stress due to Q (membrane: allowed = S; mem + bending: = 1.25S)

x_2 = width of the shell from the saddle towards center allowed for strength $= 0.78\ \sqrt{(R_m\ t)} = 91.87$

x_1 = width of the shell from saddle to tan allowed for strength = $\min(a - b/2, x_2) = 91.87$

$k = 0.1$ = if saddle welded to wear plate or shell $k = 0.1$, else $k = 1$

$A_r = A = 14210$ =resisting area for σ_6 (A for stiffener ring in-plane, $t(b + 2\ x_2)$ both sides)

σ_6 = membrane stress at the base of the saddle $= -K_5\ Q\ k/A_r = -4.825$

$K_8 = Bc\ [1 - \cos(2B)/4 + 9/4Bs\ Bc/B - 3\ (Bs/B)^2]/[2\pi\{(Bs/B)^\wedge 2 - 0.5 - \sin(2B)/(4B)\}] + B\ Bs/2\pi = 0.3405$, Where $Bs = \sin\beta$, $Bc = \cos\beta$ and $B = \beta$

$K_9 = [\{-0.5 + (\pi - \beta)\cot\beta\}\cos(\rho) + \rho\sin(\rho)]/2\pi = 0.2711$

$K_{10} = [\rho\sin\rho + \cos\rho\ \{1.5 + (\pi - \beta)\cot\beta\} - (\pi - \beta)/\sin\beta]/2\pi$, at angle ρ from top = 0.058

$Q/A = 63.34$

Membrane + bending stress in MPa *in the shell and stiffener due to Q*, all = 1.25S $\sigma = C_m\ Q/A + C_b\ Q\ R_m/Z$, max at horn for the in plane stiffener and at *I* and *J* (Fig. 11.1) for both sides of the stiffener

	shell	stiffener	σ is max at horn (inplane); at I, J (bothsides)
$Q R_m/Z$	893	1752	
σ	–69	114	stiffener inside and in-plane, C_m = (shell = $-k_8$, stif = $-k_6$), Cb = (shell = k_8, stif = k_6)
σ^1	35	–84	stiffener inside and both sides, C_m = (shell = $-k_9$, stif = k_{10}), C_b = (shell = k_9, stif = $-k_{10}$)
σ^1	26	–71	stiffener outside and in-plane, C_m = (shell = $-k_8$, stif = k_6), C_b = (shell = k_8, stif = $-k_6$)
σ^1	–69	119	stiffener outside and both sides, C_m = (shell = $-k_9$, stif = $-k_{10}$), C_b = (shell = k_9, stif = k_{10})

Note 1: these stresses are for reference only for clarity of signs in other types of stiffeners. All constants (K) are taken from Refs. 1 and 2.

11.1.4 Design Philosophy

Stress at the horn of the saddle will govern in most of the pressure vessels.

1. For D/t < 160, the wear plate may not be required.
2. For L/D < 4, horn stresses will increase.
3. Locate the saddle at 0.2L with an angle of 120° for normal lengths. For extra-long shells, locate the saddle at about 0.118*L* to stress closely at the saddle and in between saddle and minimum longitudinal stress for 120° saddle angle.
4. For D/t > 160, provide the wear plate of width =b+1.56√(Rm t), extension 5°, and thickness up to the twice shell thickness.
5. For reducing horn stresses, increase the angle to 165° and then increase the wear plate thickness
6. For further reduction of stresses at horn, select the stiffener ring in order: 1) inside and both sides, 2) outside and in-plane, 3) inside and in-plane, and 4) outside and both sides.

11.1.5 Saddle Part Design

11.1.5.1 Splitting Force

The notation is as per Figure 11.1 and example 11.1. Splitting force (F_h) in the vertical plane at the middle of the saddle is the horizontal component of the radial saddle reaction over the saddle arc with the shell and acts at CG of the saddle arc from the bottom of base plate (d). The value of *d* can be calculated by geometry and given by

$$d = B - R_m \sin(0.5\theta)/(0.5\theta)$$

Splitting force can be derived by integration and resisted by the area of section with web, base plate, and wear plate. The web depth is effective up to $R_m/3$ only. F_h is given by

$$F_h = \frac{Q(1+\cos\beta - 0.5\sin\beta)}{(\pi - \beta + \sin\beta\cos\beta)}$$

F_h induce tensile stress = F_h/A and moment $M = F_h d$
Bending stress = M/Z
where A = resisting area of the wear plate, web, and base plate at the vertical middle section of the saddle.

$$= b_1 t_1 + d_w t_w + b\, t_b$$

Z = combined section modulus of the base plate, web, wear plate, and shell with effective width $b_1 + 1.56\sqrt{R_m t}$ at the vertical middle section of the saddle.

11.1.5.2 Compressive and Bending Stress in Web and Rib

Compressive stress due to Q is uniform in web and rib, but longitudinal horizontal base force due to wind/seismic or friction will induce bending stress and varies with web height and is maximum at the end and minimum at middle.

For the purpose of calculation, divide the web length equal to the number of ribs, each with length equally dividing the web between ribs. Moment is calculated at web to wear plate joint where it is maximum. The detailed analysis is illustrated by example 11.3.

Example 11.3: Calculate stress in web and rib. Notation is given in Figure 11.1 and example 11.1

F = 100 KN, Q = 600 KN, L_b = 2000 mm, and b = 250,
Thickness of web and rib = $t_w = t_r$ =10
Height of inner rib = h_i =250
Height of outer rib = h_o= 600
Effective loaded length rib + web (outer & inner) $L_w = L_i = L_o$ = 500
Compressive force due to Q = Q L_w/L_b = 150KN
Area of each part is same = $L_w\, t_w + (b - t_w)t_r$ = 7400 mm²
Comp. stress f_C also same = 150*1000/7400 = 20.3 MPa
Force on each part due to F = F L_w/L_b
Moment in outer part M = (F $L_O/L_b)h_O$ = 15 KNm
Z for outer part = $t_r\, b^2/6 + (L_O - t_r)t_w^2/6$ = 112332 mm³
Bending stress in outer part fb = M/Z = 133.5 MPa
Combined stress in the outer part = $f_c + f_b$ = 153.8 MPa

Similarly, combined stress in the inner part = 76 MPa

11.1.5.3 Bending Stress in Base Plate due to Base Pressure

Depending on the arrangement (fixed or sliding), construct the loading diagram for an unsupported portion of base plate one side of web and between each pair of ribs. Assume three-side SS and fourth side free under base pressure as per the arrangement, calculate bending stress due to pressure from pedestal using flat plate theory

(Table 26.10 of Ref. 3). If the base plate is subjected to uplift due to high horizontal forces, base plate analysis is as per 11.5.1

11.2 SKIRT

Vertical vessels are normally supported by a cylindrical part called skirt shown in Figure 11.4. When the mean diameters of the skirt and shell approximately coincide as shown in (Figure 11.4b), the localized stresses are minimized. When the skirt is attached to the shell below the head tangent line (Figure 11.4a), localized stresses are induced in proportion to the component of the skirt reaction which is normal to the head surface at the point of attachment. Instead of a cylinder, a cone is used to provide more resistance to horizontal load (Figure 11.4c), and lap welded (Figure 11.4d) requires checking shear stress. In other cases, an investigation of local effects may be warranted depending on the magnitude of the loading, location of skirt attachment, etc., and an additional thickness of the vessel wall or compression rings may be necessary. When the temperature of the vessel is very high, highly localized thermal stresses are induced in the vessel and skirt at attachment, necessitating finite element analysis or hot-box arrangement. Apart from above, the design of the skirt is simple, calculate compressive stress and bending stress at the skirt bottom for all prevailing loads and compare with individual allowed as well as unity check. Local stresses in the shell can be analyzed approximately (due to oversize attachment) assuming the head as spherical with appropriate radius and applicable loads (reaction and moment) by WRC107 or by software Nozzlepro. Base plate and foundation bolt design is complicated and covered in base plate section 11.5.2.

11.3 LEG AND LUG SUPPORTS

These supports are used for small vertical or horizontal vessels, and stresses in vessels due to support reactions are analyzed as described in Chapter 8, and capacity of support to transfer vessel loads to supporting structure or ground is compiled with simple structural analysis using beam theory.

11.4 BOLTED BRACKET

Brackets with welded sections are used to support components, in which rectangular bracket plate welded or bolted to structure is subjected to moment. The plate bottom portion is under compression, the rest (top) portion is not in contact with structure, and bolts in this portion will be in tension. Neutral axis will get shifted down from the middle line. Design of the plate is covered in 11.5.1 case-4.

11.5 BASE PLATE

Base plates are rectangular shape used in saddles and a circular ring type used in the skirt. The function of support is to take weight of large equipment and transfer it to earth. Because the support material (metal) bearing strength is many times higher than that of earth, the cross-sectional area of support required for the purpose is very less compared to that of earth. Therefore, the bottom most part of support requires a

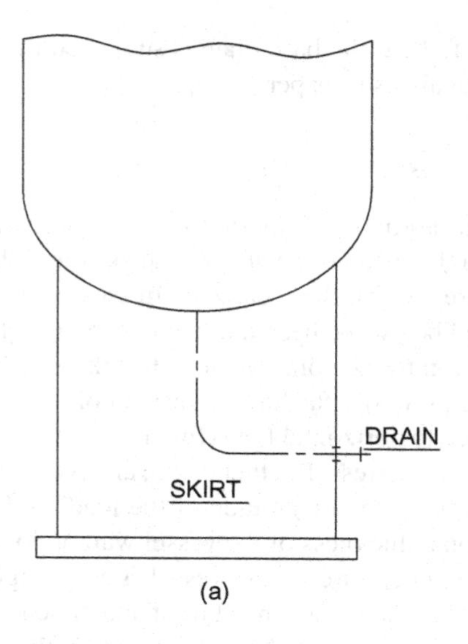

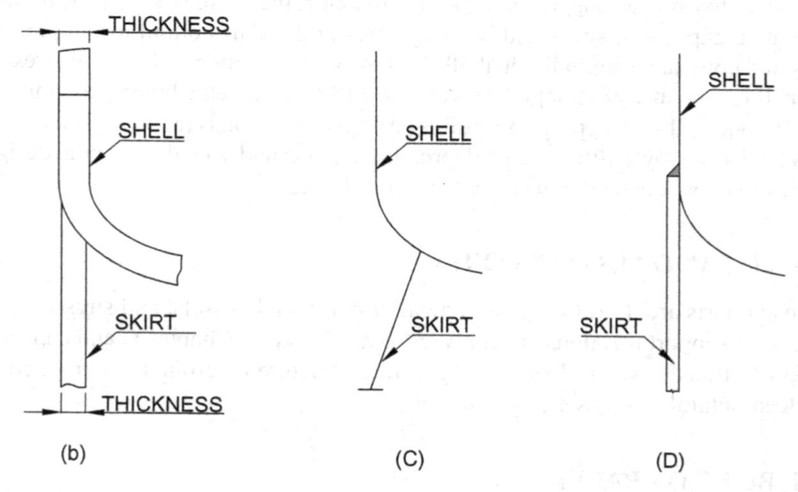

FIGURE 11.4 Skirt support.

small area of cross section to receive load and large area to transfer the same to earth. This is generally done in two or more stages. The first stage is by the base plate, which is bottom most horizontal part (plate) of support, transferring the vessel load to support base (concrete pedestal or supporting structure).

Base plates in saddles receive load through thickness of a set of plates (like rib and web plates in saddles) and transfer to support base through:

1. Directly in case of fixed support.
2. Through anti-corrosive single liner plate and anchor plate in case of sliding support with or without uplift.

3. Through low friction multiple liner plates and anchor plate in case of sliding support without uplift.
4. Through anti corrosive rollers and anchor plate in case of sliding support without uplift when a large amount of sliding in only one direction is involved.

Design of the base plate involves the calculation of thickness of base plate and bolt size and quantity for shear due to horizontal base load and tension due to uplift. Thickness is calculated from pressure by support base and/or from the free body diagram of the base plate depending on arrangement. Base plate thickness calculation is based on the following assumptions:

The base plate and pedestal surface area are plane before and after applying loads. Load received from equipment is through part of area like by web and ribs in saddle, rest of the portion will lose contact (if base plate thickness is less) with pedestal and bearing pressure on pedestal is not uniform max under ribs and web where concrete may get crushed, and no pressure where contact looses. By providing sufficient thickness (not to yield and minimize deflection), non-uniformity can be reduced.

Calculation of reaction pressure on the base plate bottom and tension in bolts is described in 11.5.1 and 11.5.2

11.5.1 Rectangular Base Plate

Refer Figure 11.5 for notation, loadings, and pressure diagram.

B = width of the base plate.

a = distance from the compression edge to tension bolt center

A = total area of bolts in tension

T_p = pretension in bolts

P = maximum pressure at the compressive edge of base plate or reaction per unit length or load per unit width and length

T = tension in bolts

R = total reaction from support base through the compressive pressure area. Because the compressive load area is triangular, R acts at k/3 distance from the compressive edge and equal to p B k/2

k = contact length

n = E ratio of steel to steel/concrete and steel to concrete = 8 to 15

There are four cases depending on direct load and moment in either direction.

Case-a: Only vertical and symmetrical load (Figure 11.5a): pressure is uniform over entire base plate bottom and equal to load/area. Pressure without load considering pretension

$$P = T_p/(BL);\ \text{with load}\ P = (W + T_p)/(BL)$$

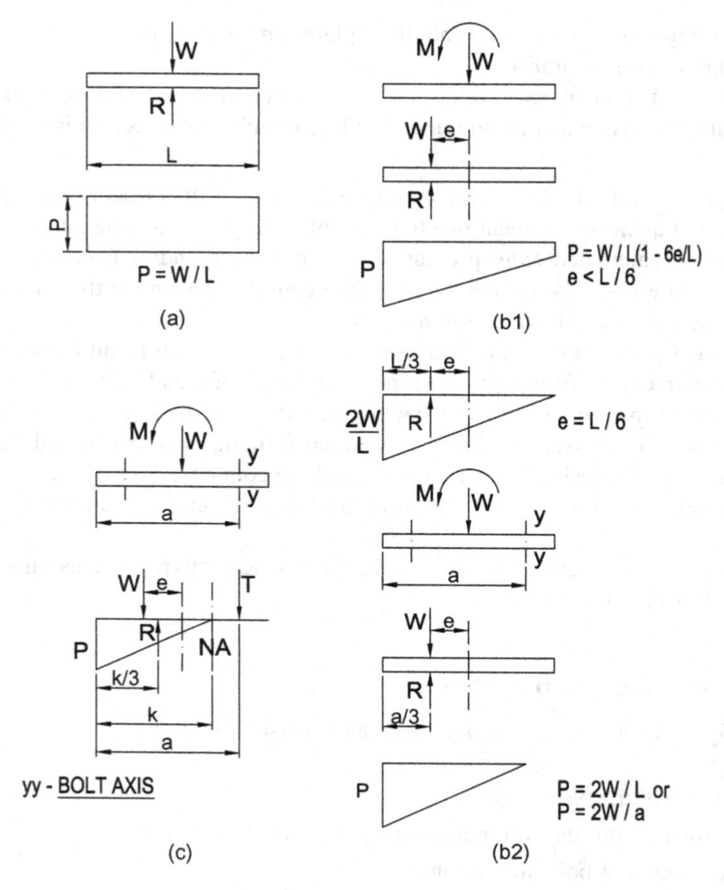

FIGURE 11.5 Rectangular base plate.

T_p is not considered, and B is suppressed for simplicity here after. *P* is given by

$$P = W/L$$

Case-b: Vertical load *W* moment *M* without uplift (Figure 11.5b): Pressure due to *M* is varying max at edge and zero at middle. *P* due to *M* as per Eq. 11.6 is

$$P = 6M/L^2 \quad (11.9)$$

Load *W* and moment *M* can be represented as *W* alone acting at an eccentricity (e = M/W) from middle as shown in Figure 11.5, which will give the same analysis results as the sum of individual analysis. Substituting $M = W\,e$ in Eq. 11.9, we obtain $P = 6W\,e/L^2$ and total *P* due to W and M is given by Eq. 11.10.

$$P = W/L(1 + 6e/L) \quad (11.10)$$

Up to $e < L/6$, the pressure P varies max at the compression edge and min at other edges. Total reaction $R = W$ *and* no uplift

For $e = L/6$, P at the compression edge is given by Eq. 11.10, substituting $e = L/6$, $P = 2W/L$ *and* zero at other edges as p due to W *and* M is equal and opposite. Uplift is zero. That is for values of $M \leq W\ L/6$, there is no uplift.

Even if $e = a/6$, the uplift is only from the bolt center to edge. $P = 2W/a$.

Case 3: Vertical loads and moments with uplift (Figure 11.4c): When $e > L/6$ ($>a/6$), the pressure is maximum at the compression edge and reduces to zero at a distance k from the left edge and lifts up rest of length. The uplift can be prevented by initial tightening of bolts to a torque or by over sizing bolts (up to a limit). The bolts in the uplift area will be in tension to press the base plate to keep it in position. T & P can be derived as described below

Equating vertical forces acting on the base plate

$$R = W + T \tag{11.11}$$

$$R = PB\,k/2 \text{ or } P = 2(W + T)/(B\,k) \tag{11.12}$$

Taking moment at reaction point (R),

$$T = W(k/3 - L/2 + e)/(a - k/3) \tag{11.13}$$

$$\text{Stress in bolt } \sigma_t = T/A \tag{11.14}$$

where A = total area of bolts in tension (assume size and number)

We have two Eqs. 11.12 and 11.13 and three unknowns R, T, and k apart from A. Compute third Eq. 11.15 by equating the ratio of elongation of bolt (e_t) and compression of plate (e_c), at left edge to ratio of distance of bolt and compression edge from NA.

$$e_t/e_c = (a - k)/k \tag{11.15}$$

Because the length relevant to elongation and compression is plate thickness and the same. Using E = stress/strain, that is, e_t/e_c = strain in bolt/strain in plate = $(\sigma_t/E_t)/(\sigma_c/E_c)$, Eq. 11.15 can be expressed as

$$(\sigma_t/\sigma_c)/(E_t/E_c) = (a - k)/k \tag{11.16}$$

Substituting $\sigma_t = T/A$ and $\sigma_C = P = 2(W + T)/(B\ k)$ from Eqs. 11.12 and 11.14 and $E_t/E_c = n$, in Eq. 11.16

$$T\,B\,k/[2n\,A(W + T)] = (a - k)/k$$

Cross multiplying $T\,B\,k^2 = 2n\,A(W + T)(a - k)$

Substituting T from Eq. 11.13,

$$B\,k^2W(k/3 - L/2 + e)/(a - k/3) = 2n\,A\,W[1 + (k/3 - L/2 + e)/(a - k/3)](a - k)$$

$$B\,k^3/3 - B\,k^2(L/2 - e)/(a - k/3) = 2nA(a - k)(a - k/3 + k/3 - L/2 + e)/(a - k/3)$$

$$B\,k^2/3 - B\,k^2(L/2 - e) = (2n\,A\,a - 2n\,A\,k)(a - L/2 + e)$$

$$B\,k^3/3 - B\,k^2(L/2 - e) + 2n\,A\,k(a - L/2 + e) - 2n\,A\,a(a - L/2 + e) = 0$$

$$k^3 - 3(L/2 - e)k^2 + [(6n\,A)(a - L/2 + e)/B]k - a(6n\,A)(a - L/2 + e)/B = 0$$

$$k^3 - 3(L/2 - e)k^2 + k_1k - ak_1 = 0 \qquad (11.17)$$

where $k_1 = 6n\,A(a - L/2 + e)/B$

Assume *k*, solve *k* from Eq. 11.17, alter *k* until Eq. 11.17 is true, substitute the *k* value in Eq. 11.13 to obtain *T*. From *T*, calculate the number of bolts required. If the number of bolts is more than assumed, increase size, alter *A*, and repeat the procedure. After verifying *A and k*, the *P* value can be obtained from Eq. 11.12. The above calculations are illustrated with examples in Table 11.1.

Thickness can be calculated depending on arrangements of support parts transferring load to the base plate, and base plates can be divided into several portions bounded by load transferring parts. Calculate the thickness of each portion using pressure load P by flat plate theory described in Chapter 10.

Case 4: For given vertical load *W*, increasing moment *M*, *e* increases. When *e* = *L/6* uplift starts, *T=0* up to *e= a/6*, with further increase of *e*, uplift, tension, and pressure increase. When *e* approaches infinity (W = 0), theoretically uplift is 100% only line (edge) contact exists.

Practically, the compression edge will compress until yield (Y) and form area required to resist bolt tension. The contact area with small length reaches yield stress and rest remains elastic as with vertical load. Neglect elastic load get *T* = *R* = *Y*area* = *M/L* and Contact width = (M/L)/(Y B)

This type of plate and loading condition may not exist. Vertical base plate of bracket support which carry only moment load with no perpendicular (horizontal) force also have bolt pretension which will act as force to form a finite value of *e* and analysis is the same as case 3.

If required alter B and repeat procedure

11.5.2 Circular Ring Type Base Plate

Base rings with the loading diagram and notation are shown in Figure 11.6 and Table 11.1. The basics of analysis are the same as the rectangular type. However due

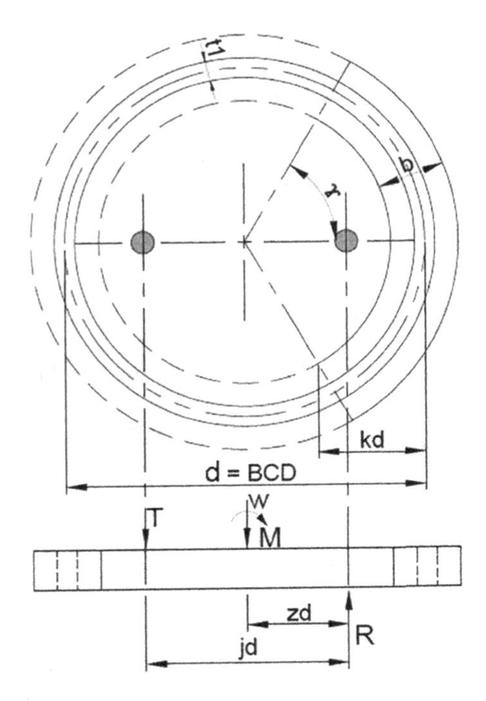

FIGURE 11.6 Circular base ring.

TABLE 11.1
Calculation of bolt stress and pressure on a rectangular base plate

For e > a/6, pretension in bolt = 0, units: N, mm (Figure 11.5)

W	1000	1000	1000	Axial load at center	
M	150000	300000	4E6	Moment at center	
e	150	300	4000	M/W > a/6, uplift	
B	1	1	1	Width	
L	600	600	600	Length	
a	550	550	550	Tension bolt to compression edge distance	
n	1	1	1	Es/Ec	
A	10	10	10	Area of each bolt, 2xM20 in tension	
k	460.18	221.35	101.092	Assume contact length = f(L, A, e, n, a, B)	
k1	24000	33000	255000	k1=[6A.n(a – L/2 + e)/B]	
k'	0.94	–0.45	–9.45	$(k^3 - 3(L/2 - e)k^2 + k1\ k - k1\ a)/100$,	
a/6	91.66667	91.667	91.6667	Alter k until k' is almost 0	
T	8.555916	154.94	7231.61	W(k/3 – L/2 + e)/(a – k/3)	
fb	0.427796	7.7469	361.58	Stress in bolts = T/2A, all. = 172MPa	
If required alter A and repeat procedure					
P	4.383311	10.435	162.854	2(W + T)/(B k)	If required alter B and repeat procedure

to the circular shape and not solid plate, the bearing width *b* and area vary along any axis, and the resisting pressure diagram for moment load is hyperbolic max at edge and zero at neutral axis (NA) in case of uplift. Tension in bolts also varies max in the outer pair of bolts and min in near pair to NA. Due to the above difference in the circular ring from rectangular plate, the equation for max compressive pressure force at edge due to moment can be derived similar to Eq. 11.6 as

$$P = 4M/d$$

The uplift condition can be compiled by the equation

$$W - P = W - 4M/d = 0, \text{ or } M/W = e \geq D/4$$

For $e < d/4$, max compressive stress = W/A + M/Z
where

A = total contact area of the base ring with concrete pedestal
$Z = \pi(D_O^4 - D_i^4)/(32D_O)$ section modulus of the base ring through axis
D_O and D_i = OD and ID of the base ring

In the uplift condition, the integration method is used for analysis by the neutral axis (NA) shift method. The bolt root area is considered as the concentric ring of thickness (t_1) with bolt circle *d* as the mean diameter for the purpose of analysis and t_1 can be computed as

t_1 = total area of bolts/(πd) = N a/(πd), where a = root area of each bolt.

The location of NA, resultant points of resisting bolt tension (T), and compressive force (R) are fractions of bolt circle diameter (d), and constant of proportions are calculated in terms of the half angle of contact arc (α) by geometry and expressed by equations below.

The contact axial width is expressed = k d, where $k = (1 - \cos\alpha)/2$ by geometry.

Location of R from center = z d
Location of T from R = j d

where

$$z = 0.5\left[\cos\alpha + C/(\sin\alpha - \alpha\cos\alpha)\right] \quad (11.18)$$

$$j = 0.5C_1/[\sin\alpha + \cos\alpha(\pi - \alpha)] + 0.5C/(\sin\alpha - \alpha\cos\alpha) \quad (11.19)$$

where

$$C = \alpha/2 - 1.5\sin\alpha\cos\alpha + \alpha\cos^2\alpha$$

$$C_1 = (\pi - \alpha)(0.5 + \cos^2 \alpha) + 1.5 \sin \alpha \cos \alpha$$

T and R are calculated by theory of equilibrium by taking moments at *R* as shown in the figure

$$T = w(e - z\,d)/(j\,d) \quad (11.20)$$

And equating forces, $R = T + W$

The geometrical area of resistances to T is equal to $t_1 d(\pi - \alpha)$ and that to R is equal to $(b - t_1 + n\,t_1)d\,\alpha$, but effective areas are less because areas nearer to NA are less effective and derived by integration.

Arc angles $2(\pi - \alpha)$ & 2α in the above equations are replaced by C_t and C_c which are functions of α. The C_c value is between α and 2α, min nearer to NA and max nearer to edge.

$$C_c = 2(\sin \alpha - \alpha)/(1 - \cos \alpha) \quad (11.21)$$

$$C_t = 2[\text{Sin}\,\alpha - (\pi - \alpha)\cos \alpha]/(1 + \cos \alpha) \quad (11.22)$$

Using C_t and C_c, stress in bolts and concrete are expressed as

$$f_s = 2T/(t_1 d\ C_t)$$
$$f_c = 2R/[(t_2 + n\ t_1)d\ C_c$$

where

t_2 = width of base ring less equivalent thick of bolts area = $b - t_1$

$n = E_s/E_c$, where E_s & E_c are Elastic modulus of steel and concrete.

Normally bolt areas required for loads with $e < d/2$ is less than to resist base shear.

First select bolts to resist shear, then calculate ring width b to resist the load W or by the empirical equation.

$$b = [W + (C_t f_s - C_c f_c n)t_1\,r]/(C_c f_c\,r)$$

Calculations are shown in Table 11.2 for various uplift cases ($e > d/4$ to max till the bolt load is equal to allowed). Bolt quantity and width b can be increased if the stresses are more than allowed for higher values of e.

Simplified method: Assuming NA at middle and considering full initial bolt tension (S a), total tension in bolts and compressive stress in concrete can be compiled as

$$T = 4M/d - W$$
$$f_C = (a\,S + W)/A + M/Z$$

where S = allowable stress of bolt

TABLE 11.2
Calculation of T and R on the base ring of skirts and stresses

Data units: mm, N, bolt all stress S = 172 MPa

W	10000	10000	10000	10000	10000	Load
M	2.51	4.11	5	10	70	Moment in KNm
e	251	411	500	1000	7000	M/W
n	10	10	10	10	10	Es/Ec
d	1000	1000	1000	1000	1000	Bolt circle dia
Preliminary selections b, a, and k						
b	150	150	150	150	150	Width of ring
	16	16	16	16	16	Bolt size
N	8	8	8	8	8	No of bolts
a	139	139	139	139	139	Area/bolt
k	0.998	0.5	0.316	0.137	0.0868	Assume
Calculation of constants f(α), alter k until k = ka						
Cos α	–1	0	0.368	0.726	0.8264	1 – 2k
Sin α	0.092	1	0.93	0.688	0.5631	
α.rad	3.05	1.571	1.194	0.758	0.5981	A cos (1 – 2k)
α.deg	174.7	90	68.41	43.45	34.269	
Z	0.251	0.393	0.434	0.472	0.4825	Eq. 11.18
J	0.751	0.785	0.782	0.77	0.7647	Eq. 11.19
Cc	3.135	2	1.552	1.001	0.7927	Eq. 11.21
Ct	0.123	2	2.407	2.802	2.9183	Eq. 11.22
Calculation of T, R, fs, fc & ka; alter k till k = ka						
T	0.034	233	842	6854	85229	w(e – zd)/(jd)
t1	0.354	0.354	0.354	0.354	0.3539	N a/(πd)
fs	0.002	0.658	1.977	13.82	165.04	2T/(t1 d Ct)
If required reselect *a* and repeat procedure						
R	10000	10233	10842	16854	95229	T + W
t2	149.6	149.6	149.6	149.6	149.65	b – t1
fc	0.042	0.067	0.091	0.22	1.5685	2R/[(t2 + n t1)d Cc]
If required reselect b and repeat procedure						
ka	0.996	0.504	0.316	0.137	0.0868	1/{1 + fs/(n fc)}
Simplified equations:						
T	40	6440	10000	30000	270000	4M/d – W
fc	0.093	0.107	0.114	0.157	0.6665	(a S + W)/A + M/Z

A = total contact area of ring = πd b, Z = (π/4)d²b

11.5.3 Base Ring Thickness

Base ring thickness can be calculated by flat plate theory. The plain ring shown in Figure 11.6 can be simplified by considering as the cantilever plate beam with unit width, fixed at skirt joint, skirt joint to OD of ring (L) as span and UDL equal to pressure (f_c) per unit width. It is used for small vessels and short spans.

Gussets are added at pitch (p) for larger vessels as shown in Figure 11.7b to reduce thickness. Analysis is by considering as a rectangular flat plate between gussets with three-side supported and fourth side free and pressure load f_C N/mm². If p > 5L, it is the same as cantilever like a plain ring. If closely pitched calculate as case 2 (SS) or

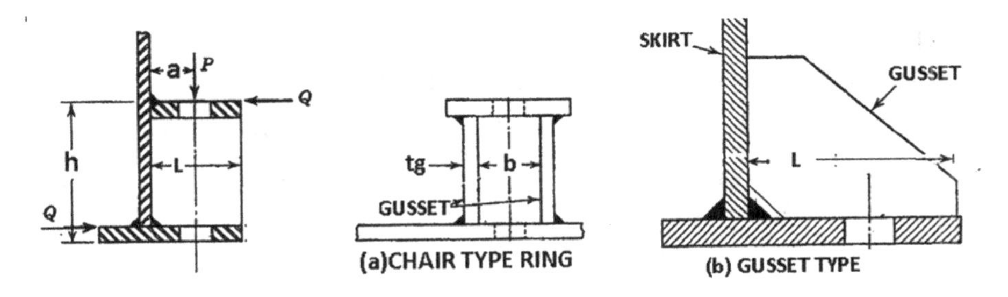

FIGURE 11.7 Base ring reinforcement.

10 (fixed) of Table 26 in Ref. 3. Compressive stress in gusset is calculated as the plate column under load (f_C /cos α) p L. The resisting area is (L cos α tg).

$$\text{Stress} = f_C\, p / \left(t_g \cos^2 \alpha\right)$$

where t_g = thickness of gusset and α = angle gusset make with vertical

The chair (Figure a) or full ring type is used in tall vessels with large *M*/*W* like chimneys to provide reverse moment at the base ring joint as shown in the figure to ensure no rotation at bottom. Ring plate calculation is the same as the ring with gusset. The top plate (compression ring) is ideally flat plate three-sides SS and with central bolt load (P) over area. There can be several assumptions to use available formulas in Table 26 of Ref. 3. To simplify assuming as SS beam between gussets with concentrated load at middle equal to bolt tension *P*. Rectangular gusset is plate under compression (neglect bending from the top plate) load *P* as shown in Figure 11.7.

Reaction in the skirt due to chair: The chair produces reaction (Q) in skirt at its joint at top ring.

Q = P a/h where P = bolt load, a and h are as shown in the figure.

The empirical equation for the thickness of the skirt due to Q is

$$t = K(Q/S)^{2/3}\, R^{1/3}$$

where,

$$K = \sqrt{3}/(1-v^2)^{\frac{1}{6}} = 1.76 \text{ for v} = 0.3$$

R = outside radius of skirt

where S = allowed stress

Guidelines to selection of parameters:

No of chairs = (4, 8, 12, 16, 20, 24). For OD of skirt in m (<1, 1.5, 2, 2.5, 3, >3)

Chair proportions: d = nominal diameter of bolt,

TABLE 11.3
Bolt load capacities in KN as per IS-800 and IS-1367

Bolt dia	Shear capacity of bolt		Tension capacity of bolt	
M-series	CL 4.6	CL 8.8	CL 4.6	CL 8.8
10	6.06	12.11	9.44	18.88
12	8.72	17.44	13.59	27.19
16	16.95	33.9	26.42	52.85
20	26.48	52.97	41.28	82.57
24	38.14	76.27	59.45	118.9
30	61.13	122.26	95.29	190.58
33	78.14	156.28	121.81	243.62
36	90.95	181.89	141.77	283.54
40	105.94	211.88	165.14	330.28

Distance between gussets = d + 2*100

Bolt hole in the base ring shall be higher than in the top plate = d + 25

Radial width of the ring beyond skirt = d + 2*20 or 3d min.

11.6 FOUNDATION BOLTS

Foundation bolts are subjected to shear and tensile forces. Size and quantity are calculated taking shear and tensile capacity of the bolt from the relevant standards; extracts of which from IS1137 are listed in Table 11.3.

11.7 ROLLERS FOR SLIDING SUPPORTS

To reduce friction in sliding, a saddle roller is used. The roller is situated between the A-saddle base plate and B-anchor plate and its axis is perpendicular to the sliding movement as shown in Figure 11.8. The roller is subjected to compressive

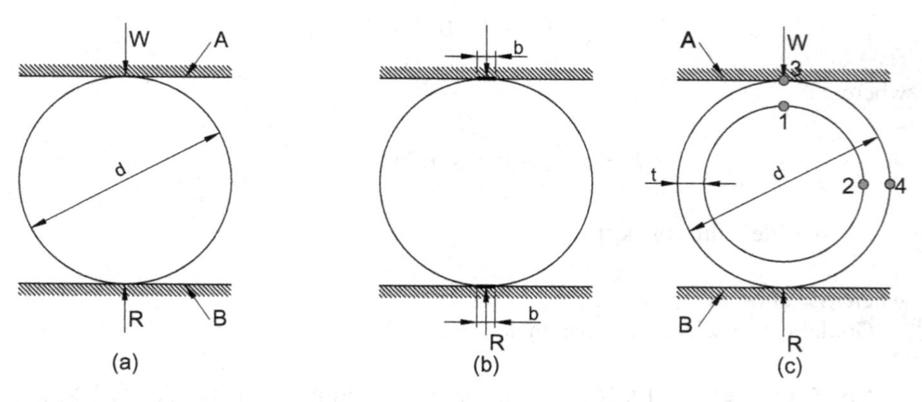

FIGURE 11.8 Roller supports.

load. Its contact with A and B is theoretically in line when there is no load as shown in (a). As load increases, the contact width increases as shown in (b). The deformation shall be within the elastic limit to facilitate smooth rolling. Calculation of load carrying capacity of the roller is complicated due to deformation of A and B. The load case 2, table 27 pressure on elastic bodies in Ref. 3 can be referred for calculating induced stress. A simple solution is possible assuming that A and B are rigid by basic elastic theory. The second method is by American Railway Engineers Association (AREA), in which an empirical formula is compiled based on practical analysis. The above methods are illustrated in example 11.4

Example 11.4: Calculate load carrying capacity of the solid roller of diameter 50 mm and length 610mm, material AISI304 having yield strength 205MPa and Young's modulus 195000 MPa.

Method 1: d = 50, L = 610, Y = 205, and E =195000

Sc = allowable compressive stress = 0.6Y= 123
e = allowable strain = S_C /E = 0.000631
c = compression on top or bottom = e d/2 = 0.1577 mm
b = corresponding contact width by geometry = $2\sqrt{c(d-c)}$ = 1.77 mm
W = load carrying capacity of roller = area x Sc = b L S_C = 131041 N

Method 2 (American Railway engineers association formula): data is same as above d = 2″, L = 24″, Y = 29733 psi

p = allowed load lbs per inch length = A d(Y – 13000)/20000 = 1004
A = if(d < 25,600,3000) = 600
W = p L = 24096 lbs, or 107442 N

Difference in *W* for both methods is due to assumptions in method 1.

A pipe can be used in place of a solid roller as shown in Figure 11.6c. Analysis depends on the d/t ratio for load per unit length *W*.

For thick pipes d/t less than 7, stresses (+ tensile, – compressive) are direct and the same at points 1 & 4(+), 2 (–), but varying at point of application 3, which is (+) and changes to (–) at d/t = 2.5. Max stress is at point-2 for values of *d/t* less than 3, and at point-1 for *d/t* more than 3. The stress equation valid for *d/t* values 3 to 7 is given by

$$\frac{kW}{\pi d/4}$$

where k depends on d/t and is equal to 5.3 for d/t = 2.5, 6.16 for d/t = 3, and as per Eq. 11.23.

$$k = -0.0687 + 4.903d/t - 0.6846(d/t)^2 + 0.1173547(d/t)^3 \quad (11.23)$$

For thin pipes with d/t greater than 7, the pipe is unstable and bending stresses are induced. Stress calculation is based on curved beam theory. Normally, a pipe is used with *d/t* values 3 to 7.

Pipe analysis with the above diameter and material, and 10mm thick: d/t = 50/10 = 5, k = 16

$$\text{Stress} = 16(104058/610)/(3.14 \times 50/4) = 34.5\ \text{MPa}$$

11.8 HANDLING AND TRANSPORT

Attachments are required to handle the equipment and its parts during manufacturing, loading and unloading for transport, erection, and maintenance. External parts such as cranes, shackles, ropes (sling), and spreader beams are used for the purpose. Attachments can be several to designers expertise. Normally, lifting lugs, trunnions, and brackets are used. Attachments induce local stresses in equipment. Lifting arrangement involving the above parts can be several to designers structural basics and exposures to the equipment. It shall be simple, safe, and economical. The first step is to select the arrangement. Variables are weight and geometry. The following guidelines are useful in designing the arrangement.

1. Light weight and short (compact) vessels are lifted by welding single lug at top passing through its CG. Heavy and longer vessels are lifted by two cranes and four lugs. If lugs are not suitable, use trunnions. Spreader beams are used if needed to reduce load on lugs. The Table 11.4 gives suggested arrangements for weights and lengths of vessels. For thin vessels pads are used below the lug or trunnion to reduce local stress. Furthermore, thin vessels can be lifted without lugs by simply wrapping the rope around the vessel.
2. Location of the lug shall be above the CG of vessel and such that the max load is transferred to lug through min number of parts if not directly, preferably avoiding bending in any part.
3. Lifting involves impact and factor 1.5 to 2 is used.
4. The sling inclination (θ-sling angle) with vertical shall be as less as possible to reduce tension in sling. Lesser the angle longer the sling required. Therefore, the angle shall be between 30 and 45°, max 60°.
5. The loads on lug shall be longitudinal (F_L) preferably which induce tension in lug, or in two directions, transverse (F_t) and normal (F_n) which induces bending. Normal load shall be avoided as bending resistance of lug is poor. Provide stiffeners if normal load cannot be avoided.

TABLE 11.4
Lifting arrangement of vessels for loads and lengths

	Short	Medium	Long
Light	Single lug	1-crane, 2-lugs	1-crane, 2-lugs
Medium	1-crane, 2-lugs	1-crane, 2-lugs	2-crane, 4-lugs
Heavy	1-crane, 2-lugs	2-crane, 4-lugs	2-crane, 4-lugs

11.8.1 Analysis

The analysis involves the following

1. Similar to any attachment on vessel, lifting arrangement is the same but in reverse direction. That is, the vessel is suspended from lugs. Loads are gravity loads with impact. Reactions can be in three directions at lug locations. All the parts involved in transferring the loads to lug are to be checked for its capacity to resist the relevant forces.
2. Local stresses in equipment part to which the lug is attached. For the purpose of analysis using WRC107, plane lug is taken as a rectangular attachment. A lug with stiffeners is equal to rectangle enclosing lug and stiffeners. Refer Chapter 8 for analysis.
3. Load distribution and stresses in lug: Let the tension in sling is *T*. Depending on the sling direction in relation to lug long axis, *T* is resolved in three principal directions of lug to get F_L long, F_t transverse and F_n normal loads as shown in Figure 11.9.

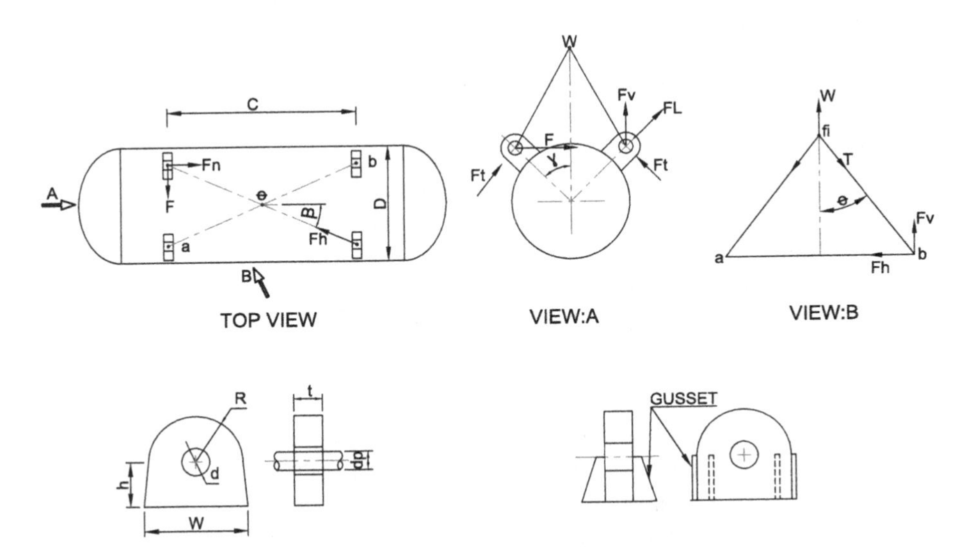

FIGURE 11.9 Horizontal vessel lifting arrangement.

The analysis of lifting attachments is simple. Compilation of forces is by static equilibrium. Stresses in attachment parts are compiled by elastic and beam theory and simple except the stresses at hole above of lifting lug. Due to shackle pin reaction in hole, the lug is subjected to stresses given below (ref: ASME BTH-1-2008). Notation as per Figure 11.9.

Tensile stress = T/[t min(2b,8t)]

Double plane shear stress = max(F_L, F_t)/(2t A)

Single plane fracture stress = max(F_L, F_t)/(t B)

where

dp = diameter of shackle pin

b = r – d/2

A = b + dp/2{1 – cos(55d_p/d)}

B = b{1.13 + 0.92/(1 + b/d)}

Compilation of forces and stresses in lifting lug for a horizontal vessel as shown in Figure 11.9 is illustrated in example 11.5. Lug dimensions are suggested as follows:

Hole dia d = $\sqrt{}$(T/100), T in N

Radius of lug r = 7d/3

Width of lug at bottom *w* depends on forces, for *FL* straight, for *Ft* and *Fn* taper max 30°

Height h = min 1.25d, max no limit for *FL*, 1.5d for *Ft* and *Fn*.

Weld strength shall be for shear or tension applicable for fillet or partial welds.

Example 11.5: Calculate stresses in lugs on horizontal vessel Units: mm, N, and MPa

θ = Angle, lifting sling make with vertical = 45°

α = Lug longitudinal axis (radial) angle with vertical = 30°

D = Outside diameter of the vessel = 1756

C = Distance between the pair of lugs along shell long axis = 2700

L = Load of equipment on four lugs to lift = 200000 N

n = Number of lugs = 4

F = Factor of safety (impact factor) = 2

β = Angle as shown in figure =A tan [C/(D sin α)] = 72°

Calculation of forces

W = Design load per lug = L F /n = 200000*2/4 = 100000

T = Design load per lug = L F /n = 200000*2/4 = 141421

F_h = Horizontal component of T = T sin θ = 141421/sin 45 = 100000

F_V = Vertical component of T = T cos θ = 141421/cos 45 = 100000

F_n = Component of Fh normal to lug = F_h sin β = 100000 sin 72 = 95097

F = Hor. component of $F_h = F_h \cos\beta = 100000 \cos 72 = 30624$

F_L = Longitudinal load on lug = $F_v \cos\alpha - F \sin\alpha = 100000 \cos 30 - 30924.2 \sin 30 = 71140$

F_t = Tangential load on lug = $F_v \sin\alpha - F \cos\alpha = 100000 \sin 30 - 30924.2 \cos 30 = 76781$

F_b = Bearing load between shackle pin and lug = $\sqrt{F_L^2 + F_t^2} = 104671$

Lug data

d = Diameter of hole [$\sqrt{}$(T/100)] = 36
r = Radius of lug at hole center(min.7d/3) = 54
w = Width of lug at bottom = 105
h = Height of hole from bottom = 54
t = Thickness of lug = 40
Y = Yield stress of lug material (IS2062) = 240

Stresses at hole section & above

b = Ligament of lug = r – d/2 = 36
A = b + d_p/2{1 – cos(55d_p/d)}, (assume d_p diameter of shackle pin = d/2), = 51.43
B = b{1.13 + 0.92/(1 + b/d)} = 57.24
σ_t = Tensile stress = F_L /[t.min(2b,8t)], (all = 0.6Y = 144) = 24.7
σ_s = double plane shear stress max(F_L, F_t)/(2t A), (all = 0.45Y = 108) = 25.93
σ_f = Single plane fracture stress = max(F_L, F_t)/(t B), (all = 0.6Y = 144) = 17.95
σ_b = Bearing stress = F_b/(d t), (all = 1.25Y = 300)

Stresses at lug bottom

σ_t = Tensile stress = F_L/(w t), (all = 0.6Y = 144) = 16.94
σ_{bt} = Bending stress = F_t h/(t w^2/6), (all = 0.66Y = 158) = 56.41
σ_{bn} = Bending stress = F_n h/(w t^2/6), (all = 0.66Y = 158) = 183.4

Combined stress ratio = σ_t/0.6y + (σ_{bt} + σ_{bn})/0.66Y, (all = 1) = 1.632

Note: σbn & unity check is exceeding th allowed, gussets shall be provided as shown in figure.

11.8.2 Transport

Transportation loads are called acceleration loads. Their structural effect is acceleration in three directions. Vertical acceleration is due to bad roads and water waves, longitudinal is due to acceleration, and braking and transverse are due to approaching curves. Recommended accelerations are

For surface transport: Vertical 1.5, longitudinal 1, and transverse 0.5

For sea transport: Vertical 1.5, longitudinal 1.5, and transverse 1

Effect in vessels: Effects depends on transport arrangements made to resist movements. Normally, saddle base plates are tied to vehicles but the vessel above the saddle is not effectively prevented to transfer horizontal loadings to vehicles; as a result, saddles are subjected to moment. If the acceleration is more than that in operation due to wind/seismic, the saddle will deflect. Therefore, the saddle shall be checked for the above acceleration loads.

REFERENCES

1. Code ASME S VIII D 1, 2019.
2. *Process equipment design*, L. E. Brownell and E. H. Young, 1959.
3. *Formulas for stress and strain*, Reymond J. Roark and Warren C. Young, 5th edition.

12 Wind and Seismic

12.1 WIND

Wind can be described as a highly turbulent flow of air sweeping over earth's surface with variable velocity, in gusts rather than in a steady flow. The direction is usually horizontal. It may contain vertical components when passing through hills/valleys. Velocity increases up to a certain elevation and remains constant above that elevation.

The structural effect of wind velocity is pressure (P) equivalent of total kinetic energy of air mass on a flat surface perpendicular to wind velocity and is given by

$$P = \rho V^2/2g = 1.29/(2 * 9.81)V^2 = 0.066\,V^2\ \text{kg/m}^2 \text{ or}$$
$$P = \rho V^2/2 = 1.29/2\,V^2 = 0.645\,V^2\ \text{N/m}^2$$

where

ρ = density of air under ambient conditions 1.29 kg/m^3
g = gravitational acceleration 9.81 m/s^2 (32 fps^2)
V = basic wind speed (Velocity) in m/s for various zones of countries in their standards

The Indian standard (IS: 875 Part 3) divides India into *six zones* on basic wind speed, lowest for plains at lower elevations and highest for hill areas, and values are 33, 39, 44, 47, 50, and 55 m/s respectively.

There are three multiplying factors to obtain design wind speed.

1. Risk coefficient $k_{1.}$
2. Terrain factor $k_{2.}$
3. Topography factor $k_{3,}$

The risk coefficient is given by the equation

$$k_1 = \frac{A - B\,loge(1/N)}{A + 4B}$$

where

A = {83.2, 84, 88, 88.8, 90.8} for six zones.
B = {9.2, 14, 18, 20.5, 22.8, 27.3} for six zones.
N - Design life (5, 25, 50, 100) years.

Terrain factor k_2 depends on the category (1 to 4), class (A, B, C), and height (h) of structure (10 to 550m). It is minimum 0.67 for (cat. 4, cl. C & h = 10 m) and maximum 1.4 for (cat.1, cl. C and h = 10 m).

DOI: 10.1201/9781003091806-12

Topography factor k_3 depends on slope minimum 1 for slope < 3°and maximum 1.36 for slope > 3

Design wind speed, $V_z = k_1\ k_2\ k_3 V$

Design wind pressure, $P_z = 0.066\ V_z^2\ kg/m^2$ or $0.645\ V_z^2\ N/m^2$

Force on any vessel with a projected area of the vessel perpendicular to the wind direction (A) due to P_z is given by

$$F = P_z A$$

The shape of the equipment influences the resulting force, and the cylindrical shape reduces the force by 0.7 provided there are not many attachments and nozzles. IS 875 gives shape factors for other equipments.

This force is distributed over the vessel surface. Design (calculation of shear force, bending moment, deflection, and bending stresses) of vessels with a uniform cross section is the same as a beam with uniformly distributed load (UDL). If the vessel cross-section varies over the entire length, the UDL will vary for each length.

As wind flow vibration is induced, analysis is not required if the natural frequency of the vessel is more than 4. Up to 80m height, analysis may be carried out in only the first mode.

12.2 EARTHQUAKE

Seismic forces on a vessel result from sudden vibratory motion with horizontal acceleration (S_a) of the ground, on which the vessel is supported, and vessel response to this motion. The principal factors in the damage to structures are the intensity and duration of the earthquake motion. The ratio of this acceleration and gravitational acceleration is called the spectral acceleration coefficient and expressed by basic equation S_a/g.

The value of S_a varies depending on various factors including the location. Indian standard IS-1893(1 and 4) gives the following factors and procedure for computing horizontal seismic coefficient.

Zone factor: India is divided into four zones: II- middle, III- middle and coastal, IV-Himalayas, and V-Kutch and certain high altitudes, and zone factors (Z) are 0.1, 0.16, 0.24, and 0.36, respectively.

Category of structure: Four categories are specified

1. Life, property, and population.
2. Fire hazard/damage in plants.
3. Expensive but not serious.
4. Others.

Analysis method: Two methods are specified.

- DBE - design basis earthquake for category 1.
- MCE - max considered earthquake for other categories.

For vessels, the DBE method is used.

Other factors:

R - Response reduction factor: given in Table 9 for different structures, and two for steel stacks.

I - Importance factor: for different structures it is given in Table 8, 1.5 for steel stacks and silos and 1.75 for vessels.

S_a/g - spectral acceleration coefficient as per Figure 2, Annex B, damping factor (for steel = 0.02), and T time period ($T = 0.085H^{0.75}$ for steel frame buildings). For vessels, $S_a/g = 1.5$.

The horizontal seismic coefficient (A_h) for the DBE method is given by

$$A_h = \frac{Z}{2}\frac{I}{R}\frac{S_a}{g}$$

For the MCE method, the above equation with $Z/2 = 1$ and Sa/g as per site specific spectra is applicable.

The horizontal seismic forces acting on the vessel with total weight W are reduced to the equivalent static forces. The structure is designed to withstand a certain minimum horizontal base shear ($V_b = A_h W$) applied at the base of the vessel in any direction. The problem is how to resolve this base shear into equivalent static forces throughout the height of the vessel in order to determine shear force and bending moment in the structure at different elevations and overturning moment at base. The result depends in large part on the dynamic response of the structure, which may be assumed either rigid or flexible.

The vertical component also exists but to simplify the design procedure the vertical component of the earthquake motion is usually neglected on the assumption that the ordinary structures possess enough excess strength in the vertical direction.

12.2.1 Rigid-Structure Approach (RSA)

The rigid-structure approach is used for short heavy vertical or horizontal vessels. Resolution of base shear is simple. The structural effect of motion on the vessel is horizontal force (F) at its centre of gravity and equal to A_h W. For any elemental length, F is A_h multiplied by its weight at its CG. For the vertical vessel, it is a cantilever beam with load along its length proportional to its weight or A_h times its self-weight. Calculation of shear and bending moment at any section is like any cantilever beam. However, it cannot be applied reasonably to tall, slender columns and chimneys regardless of their dynamic properties.

12.2.2 Flexible Tall Structure Approach

The sudden erratic shift during an earthquake of the foundation of a flexible tall cylindrical vessel relative to the centre of gravity causes the vessel to deflect because the

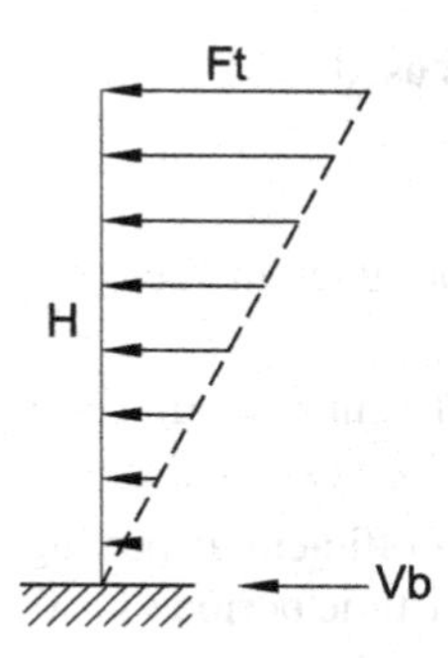

FIGURE 12.1 Triangular seismic load.

inertia of the vessel mass restrains the vessel from moving simultaneously with its foundation. Due to this, the loading is not as in the rigid vessel but zero at base, and increase along its height maximum at top with the same base shear as RSA. For the vessel with uniform weight, loading is triangular as shown in Figure 12.1. Further calculation is a simple cantilever with triangular loading. Because base shear remains the same and is equal to area of the triangle, force at top is given by

$$F_t = V_b/(H/2) \text{ N/m}$$

Shear force and bending moment at any section can be calculated from the moment area method (Chapter 3).

Shear force = area of the loading diagram from top to the section

Moment = moment of the above area at the section = shear force x distance of the section from CG of area.

Example 12.1: Given $A_h = 0.15$, W = 100 KN, H = 30 m, D = 5 m, and calculate base moment. F_o defined in the next section is assumed as zero.

Base shear V = 0.15*100 = 15 KN.

Force at top = 15/(H/2) =1KN/m.

Base moment = 15 × 2H/3 = 300 KNm.

If the vessel is non-uniform, resolving base shear is based on the ratio of elemental weight moment at the base to sum of all weight moments at the base. It is given by,

F_n = Force at the bottom of the nth section from top = $(V - F_o)w_n h_n/\Sigma(w h)$

$F_o = 0.004V(H/D)^2$, maximum $0.15V$, is a portion of V assumed concentrated at the top of the structure to approximate the influence of higher modes.

w_n = weight of the nth section.

h_n = height of CG of the nth section from the base.

The loading diagram can be built up by applying these loads at all sections. Once the loading is resolved, shear force and bending moment at any section can be calculated like any cantilever beam.

Shear force at any point = sum of loading from top to that point.

Moment at bottom of any section = moment at its top + shear force at top x height of section + A_h x section weight x height of its CG from bottom.

12.3 OTHER STANDARDS

Structural effects of wind and seismic forces are explained in 12.1 and 12.2 including the computation of wind pressure and seismic base shear as per the Indian standard. In this section, computations of wind pressure and seismic base shear are explained as per UBC, Australian/New Zeeland, and American standards.

12.3.1 UBC-1997

12.3.1.1 Wind Pressure Calculations

V = basic wind speed.

C_e = comb height, exposure and gust factor coefficient (Tab.16-G), 0.62 to 2.34.

C_q = pressure coefficient for the type and geometry of equipment. Tab.16-H, 0.8 to 4, 0.8 for Round/elliptical.

I_w = importance factor Tab.16-K

q_s = wind stagnation pressure at 33' (Tab.16-F) = $0.615V^2/g$, (g in m/s^2).

P = design wind pressure = $C_e\ C_q\ q_s\ I_w$.

12.3.1.2 Seismic Base Shear Calculation

Static analysis:

Z = Zone factor (T.16-I) = (0.075, 0.15, 0.2, 0.3, 0.4) for zones (1, 2A, 2B, 3, 4)

I = Importance factor (Tab. 16-K): = 1 for Occupancy Cat 3–5 (std) and 1.25 for occ. cat 1, 2 (hazard/essential)

Soil profile type (SA - SF) (Tab.16-J)

Seismic source Type (A, B, C) (Tab.16-U), for zone-4 only

N_v = Seismic source factor (tab. 16-T), for zone-4 only

R = Structural system Coefficient (tab. 16-N) = 2.9 for stacks/vertical vessels on skirt/all other self supporting structures and = 2.2 for vessels

C_v = Seismic coefficient (tab.16-R)

C_a = Seismic coefficient (tab.16-Q)

T = Fundamental time period

W = weight of equipment

Base shear

a. UTS *design* & rigid
   Base shear $V' = 0.7C_a I.W$ for $T < 0.06$, $= C_v I W/(R T)$ for $T \geq 0.06$
   min $= 0.11C_a I W$, max $= 2.5C_a I W/R$
   For zone-4 min $= 0.8Z N_v I W/R$
b. *Allowable stress design*

$$\text{Base shear } V = V'/1.4$$

Static analysis: all in zone-1, occupancy categories 4 and 5 in zone-2, regular < 240', irregular ≤ 65'.
Dynamic analysis is required for structure height ≥ 240' or time period > 0.7

12.3.2 Australian and New Zeeland Standard AS/NZS 1170

Example 12.2: Calculation of Design Wind pressure: AS/NZS 1170.2:2011

V_r = Regional gust wind speed (Tab-3.1), V500 & Region A11 = 45 m/s
M_d = Wind directional multiplier (Tab-3.2), Region A11 & any direction = 1
M_{zcat} = Terrain/height multiplier Tab-4.1, height 20 m and Terrain category 21 = 1.08
M_s = Shielding multiplier Tab-4.3, Assumed = 1
M_t = Topographic multiplier CL-4.4, Assumed = 1
V_{des} = Design Wind speed as per CL 2.2 = $V_r M_d M_z M_s M_t$ = 48.6 m/s
ρ = Density of air as per CL 2.4.1 = 1.2 kg/m³
C_{pe} = External pressure coefficient as per Tab-5.2 (A) = 0.7
K_a = Area reduction factor as per CL 5.4.2 = 1
K_{ce} = Combination factor applied to ext press As per Tab-5.5 = 0.9
K_t = Local pressure factor as per CL 5.4.4 =1
K_p = Porous cladding reduction factor as per CL 5.4.4 = 1
C_{fig} = Aerodynamic shape factor = $C_{pe} K_a K_{ce} K_t K_p$ as per Cl 5.2 = 0.63
C_{dyn} = Dynamic response factor As per CL 2.4.1 = 1
P = Design Wind Pressure $P = 0.5\, \rho\, (V_{des})^2 C_{fig} C_{dyn}$ = 893 N/m²

Example 12.3: Calculation of the horizontal design earthquake coefficient: AS 1170.4:2007

h_x = Height of the component from the base = 15.8m
h_n = Height of the structure from the base = 0.6m
I = Importance level as per section F2 & Table F1 AS/NZS 1170.0 = 2

P = Annual probability of exceedance as per Table F2 AS/NZS 1170.0 = 1/500

K_p = Probability factor = f(P), Table 3.1, as per site the data sheet = 1

Z = Hazard factor Table 3.2 = 0.09

Site sub-soil class = C_e

EDC = Earthquake design category as per Table 2.1, f(soil, k_p, z, h_n) = cat. II

I_c = Component importance factor as per CL-8.2 = 1.5

a_c = Component amplification factor as per CL-8.2 = 1

R_c = Component ductility factor as per CL-8.2 = 2.5

$C_h(o)$ = Bracketed value of the spectral shape factor for the period of zero seconds f(soil), Table 6.4 = 1.3

k_c = Factor (= $2/h_n$ for $h_n \geq 12$; else = 0.17) = 0.17

a_x = Height amplification factor at h_x as per CL-8.2 = 1.102

a(floor) as per CL,8.2 = $k_p.z.C_h(0)$ = 0.117

Coefficient, simple method Cl 8.3 = a(floor).ax[I_c a_c /R_c] = 0.0774

Coefficient, design acceleration method Cl 8.2 = a(floor) a_x[I_c a_c /R_c] = 0.0702

12.3.3 American Standard ASCE

Example 12.4: Wind Pressure Calculation in conformity to ASCE 7-10: C-30

V = Basic wind speed from Figure 6.1 = 45 m/s.

Surface category, smooth (B, C, D) = smooth C.

D = Diameter of the equipment = 1.93 m.

h_c = Centroidal height of equipment = 1.61 m.

Z = Height of CG of equipment from the ground level = 1.91 m.

Z_g as per Table 26.9–1 (B-365.76, C-274.32, D-213.36) for category C = 274.32.

α = as per Table 26.9–1 (B-7, C-9.5, D-11.5) for category C = 9.5.

K_{Zt} = Topographic as per 26.8.1 = $(1 + k_1\ k_2\ k_3)^2$ but = 1 if site conditions and location of the structure do not meet all conditions specified.

K_d = Wind directionality factor as per Table 26.6–1 (0.95 for round tanks) = 0.95.

I = Importance factor as per Table 6.1, for area category #3 and V > 100mph = 1.15.

C_f = Force coefficient as per Figure 29.5–1, 2, 3 (0.52 for the round cross section, moderately smooth surface and for a value of h_C/D = 0.83) = 0.52.

Z_{bar} = Equivalent height of structure = 0.6Z = 1.15 m.

C = Constant as per Table 26.9-1 (B-0.3, C-0.2, D-0.15) for category C = 0.2.

L = Constant as per Table 26.9-1 (B-97.54, C-152.4, D-198.12) for category C = 152.4.

$\in_{bar}$ = Constant as per Table 26.9-1 (C-1/3, C-1/5, D-1/8) for category C = 0.2.

g_Q = Peak factor as per Cl.26.9.4 = 3.4.

g_V = Peak factor as per Cl.26.9.4 = 3.4.

β = Horizontal dimension measured normal to wind direction = 0.9 m.

n = Natural frequency of the equipment = 33 cps.

K_z = Velocity pressure coefficient as per Table 30.3-1 = $2.01(15/z_g)^{2/\alpha}$ If z < 4.6m, else $K_z = 2.01(z/z_g)^{2/\alpha}$ = 0.85.

q_z = Velocity pressure as per 29.3 – 1= $0.613\ K_z\ K_{zt}\ K_d\ V^2\ I$ = 1153 N/m^2.

L_{zbar} = Intensity of turbulence Eq. 26.9-4 = $c(10/Z_{bar})^{1/6}$ = 0.29.

L_{zbar} = Integral length scale of turbulence Eq. 26.9-9 = $L(0.1\ Z_{bar})^{\in bar}$ = 98.81.

$$Q = \text{Background response factor Eq. 26.9-8} = \sqrt{1/\left\{1+0.63\left[\frac{\beta+Z}{L_{zbar}}\right]^{0.63}\right\}} = 0.97$$

$$G = \text{Gust factor (for a rigid structure) Eq. 26.9-6} = \frac{0.925(1+1.7g_q\ Izbar\ Q}{1+1.7g_V\ Izbar},$$
max 0.8 = 0.85

Wind pressure P = $q_z\ G\ C_f$ as per 29.5-1 = 510N/m^2

Example 12.5: Seismic base shear as per ASCE7-10.13

a_p = Component amplification factor = 1 to 2.5, Tab-13.5-1 or 6-1, 1 for boilers etc = 1

R_p = Comp. response modification factor = 1 to 12, Tab-13.5-1 or 6-2, 2.5 for boilers etc = 2.5

I_p = Importance factor = 1 or 1.5, 1.5 for life safety, hazardous material, occupancy cat-IV = 1.25

S_s = MCE spectrum response acceleration Figure 22 = 0.202

Site class (A–F) for soil properties (Chapter 20), D if not known = D

F_a = Site coefficient = 0.8 to 2.5, Tab.11.4-1 f (site class, Ss) = 1.6

Z = Height of the structure of comp attachment = 5m

h = Average roof height of the structure from the base = 5m

$S_{ms} = F_a\ S_s$ = 0.3232

$S_{ds} = 2S_{ms}/3$ = 0.2155

A_h = Horizontal seismic coefficient = $0.4a_p\ S_{ds}(1+2z/h)I_p/R_p$, (min = $0.3S_{ds}\ I_p$) (max = $1.6S_{ds}\ I_p$) = 0.4*1*0.2155*(1+2*5/5)*1.25/2.5 = 0.129

12.4 APPLICATION OF WIND/SEISMIC LOAD

Wind and seismic will not occur at the same time. Because the structural effect is different for both, it is compiled separately, and the maximum effect is considered for analysis. Because this load is occasional, allowable stress is higher than for sustained loads, and the Indian standard specifies 1.33 times that for dead loads. The structural effect (saddle reaction of wind/seismic load on the horizontal vessel) is illustrated by example 12.6.

Example 12.6: Calculate the saddle reaction (Q) due to wind/seismic on the horizontal cylindrical vessel

Design data: Wind pressure (P) = 1.5KN/m^2, horizontal seismic coefficient (A_h) = 0.294, weight of the vessel with contents (W) = 500 KN, vessel OD (D) = 2m, equivalent exposed length (L) = 10m, two saddles and distance between saddles (B) = 6 m, height of CG of the vessel from the base (h) = 1.2 m, insulation thickness (T) = 100 mm, and saddle bearing angle (θ) = 120°

Calculations:

L_b = D sin60° = 1.75 m saddle length (reaction moment arm).

A_t = L(D+2T) = 22 m^2 transversal projected area to wind.

A_L = π/4(D+2T)^2 = 3.8 m^2 longitudinal projected area to wind.

F_{tw} = A_t P = 22*1.5 =33 KN transverse wind load (shape factor 0.7 neglected).

F_{LW} =A_L P = 3.8*1.5 = 5.702 KN longitudinal wind.

F_{ts} = F_{LS}= W A_h = 500*0.294 = 147 KN seismic load.

M_{tw} = F_{tw} h = 33*1.2 = 39.6 KNm moment due to $F_{tw.}$

M_{LW} = F_{LW} h = 5.702*1.2 = 6.842 KNm moment due to $F_{LW.}$

M_{ts} = M_{LS} = F_s h = 147*1.2 = 176.4 KNm moment due to $F_{S.}$

Q_t = 3/2max(M_{tw}, M_{ts})/L = 1.5*176.4/1.75 = 151.2 KN max reaction (three times the average, refer section 11.1.3) at the edge on each saddle due to $M_{t.}$

Q_L = max(M_{LW}, M_{LS})/B = 176.4/6 = 29.4 KN.

Q = max(Q_t, Q_L) = 151.2 KN max Q to be considered in saddle analysis.

13 Flanges

A flange (loose without a hub) is a special type of flat plate *circular ring* with several loads derived from pressure to facilitate removing components such as part of the pipe from the piping system and opening of end cover (*blind flange*) for process requirements. A blind flange is a removable type of flat end closure to the cylindrical part (called *pipe* in this section). Flanges can be classified as

- Integral and loose flanges
- Raised face and full face flanges
- Ring type gasket flanges
- Special flanges

Integral flanges consist of a hub as shown in Fig. 13.1 and 13.2 and are equal to cast or forged integrally with the pipe or welded thereto by such a nature that there is no relative deformation or rotation between both the flange and pipe at their joint, whereas loose flanges with or without a hub have no rigid connection to the pipe, and the method of attachment is not considered to give mechanical strength sufficient to resist relative movement or rotation. A flange without a hub can also be considered as an integral flange by providing required weld equivalent to the hub in integral flanges.

13.1 BASICS OF FLANGE ANALYSIS

A loose flange without a hub can be analyzed as a flat plate ring as shown in Figure 13.2a by equations from case (1f) and (1L) in Table 24 of Ref. 3 with certain conservative assumptions. Resistance to bending and deflection of the portion of flange outside the bolt circle is not covered in the operating condition and inside the gasket reaction is not covered in the seating condition.

The integral flange as shown in Figure 13.1 is a combination of three elements: a flat plate (same as loose flange), a hub, and a shell. Analysis by the integration method is by assuming the longitudinal strip of unit width of the shell and hub as the beam on elastic foundation, shell uniform thick, and hub varying thick. Due to discontinuity between elements, discontinuity stresses are induced in the hub in addition to longitudinal stress. Radial and tangential stresses are induced in the flat plate element as described in flat plate theory. Ref. 2 gives the equations for stresses by the integration method. In addition, Appendix A2 of the code[1] gives equations and flange constants. Figure 13.1 shows discontinuity forces and moments in integral flanges. Furthermore, the flange rotates at the gasket reaction diameter due to operating forces and is prone to leak. Hence, a flange shall be rigid enough to resist rotation apart from withstanding operational and seating forces.

DOI: 10.1201/9781003091806-13

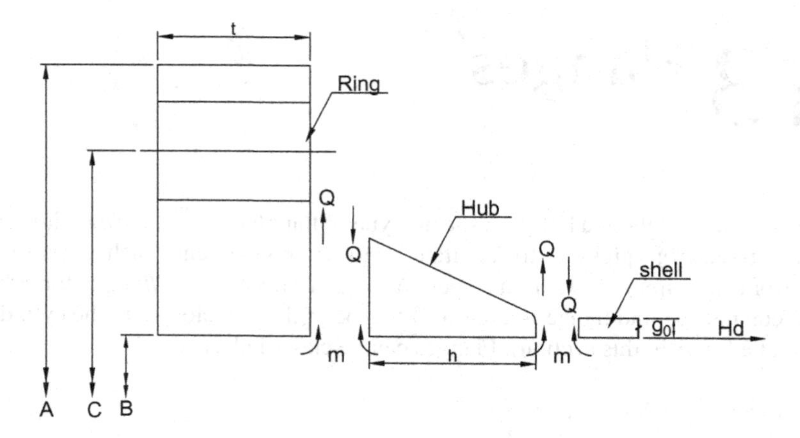

FIGURE 13.1 Discontinuity forces and moments in the integral flange.

Standards are available in most of the countries to select for pressure and temperature up to certain size. Most popular is the American standard covering up to 60" NPS, which is described in B16.5 and 16.47. For higher sizes or for non-standard sizes, flanges are designed as per the basics and code[1] and covered in this chapter.

13.2 SELECTION OF PARAMETERS AND THEIR INFLUENCE ON FLANGE DESIGN

The parameters for optimum design, leak proof, safe, and convenient maintenance are bolt size and its circ pitch, flange width, and gasket location and its parameters.

13.2.1 Circumferential Pitch of Bolts (P_c)

For analysis in the operating condition, the bolt circle diameter is considered as a support on the entire perimeter for pressure loads, which is not true. The support is only at as many as the number of bolts and at its center points. Consider the free body diagram of the sector between two adjacent bolts under the operating load whose analysis will show that the stresses are not uniform at all points on any radius. It is max in the middle. To maintain it almost uniform, the circ ligament at the bolt circle shall not be more than six times the thickness. Adding the effect of the gasket factor and bolt holding width, the equation for maximum P_C is given in various references as

$$P_C < 2a + 6t(m + 0.5)$$

where a = nominal size of bolts

t = flange thickness

m = gasket property in the operating condition

Furthermore, P_c shall be sufficient for tightening bolts with tools without obstruction from adjacent nuts. Minimum pitch to be provided is equal to twice the bolt nominal diameter plus 6 mm (2a + 6) considering standard tools. Apart from the above min and max limits, pitch depends geometrically on the quantity of bolts. Bolts shall be sufficient to withstand operating and seating loads. In case the geometrical pitch is not within max and min limits, limits can be altered primarily by the size of the bolt or by the quantity of the bolt or by the thickness of the flange or by increasing BCD in the order.

13.2.2 Flange Width

Flange width shall be minimum to reduce weight and cost. To facilitate bolt tightening, distances from the bolt circle to hub and outer edge (R and E) shall be minimum to reduce the flange width. The minimum values of *R* and *E* depend on the bolt nominal size and are given in various literature studies and TEMA against the bolt size (see extracts in Table 13.1). Larger the bolt size, larger the flange width. Therefore, the bolt size shall be minimum.

13.2.3 Gasket

For leak proof, a gasket is required between two mating flanges. The gasket shall be softer (hardness ≤ 25 BHN) than the flange material and consists of soft metal or non-metal or combination, elastically pressed between flanges when tightened by bolts to prevent leakage. Softness and material selection depend on the pressure and temperature. Higher the pressure, higher shall be the compressive strength (y) of the gasket to prevent leakage due to deformation under pressure force in the radial direction. Gaskets are procured with known properties *m* and y. m is a factor in residual stress in the gasket in the operating condition, and residual stress shall be at least two

TABLE 13.1
Values of R, E, and A_o Units: mm, mm²

Nom-dia	Pitch	R.radial	E.edge	P_c min.pitch	A_o.bolt area
16	2	28.6	20.6	38	138
20	2.5	31.8	23.8	46	217
24	3	36.51	28.58	54	313
27	3	38.1	29	60	414
30	3.5	46.1	33.34	66	503
36	4	54	39.7	78	738
42	4.5	61.9	49.2	90	1018
48	5	69	56	102	1343
56	5.5	77	64	118	1863
64	6	85	67	134	2469
72	6	90	70	150	3222
80	6	94	75	166	4077
90	6	108	85	186	5287
100	6	119	93.66	206	6652

A_o is min including negative tolerances

to three times the pressure to prevent leakage. The gasket width (N) provided is not fully effective for analysis due to the type of gasket, raised face width, and practical accuracy of alignment. The code gives the following rule for effective width (b).

For flat and raised faces, effectiveness is half of the actual width up to 12.5 mm and is less than half for higher widths and is equal to 2.5 times square root of the above reduced width. The effective width is expressed by the equation

$$b = \min\left(\frac{N}{2}, 2.5\sqrt{\frac{N}{2}}\right)$$

Table 2.5.2 of the code[1] gives the values of effective width for other arrangements.

Higher the width of gasket and its properties *m and y*, higher the gasket reaction force and bolt area required. Therefore, non-metallic gaskets are used for low pressures and for full faced. Metallic gaskets (not used for full face) filled with a material such as graphite are used for medium and high pressures normally with a narrow width from (¼" for ½" NPS to 25 mm for 36" NPS). Oval and octagonal ring type metal gaskets are used for high pressures. *m* and *y* values are very high for ring gaskets. *m* and *y* values for normally used gaskets are given in Table 13.2, which specifies the same. Spiral or concentric grooves of depth 40 microns and spacing 80 microns are used for raised face flanges. The edges of grooves will deform and hold the gasket.

The *gasket location* between the bolt and inside is sensitive. The moment lever arm for forces varies due to the location. Locating nearer to the bolt will increase the lever arm for operating forces and decrease for seating forces. Locating nearer to the inside diameter (B) will have a reverse effect. *m* and *y* values for commonly used gaskets are given in Table 13.2.

TABLE 13.2
m and y Values for Flange Gaskets

Type of gasket	m	y.MPa
Rubber	0.5	0
Rubber with fabric or asbestos fiber	1	1.5
Garlock	5.2	10
Non asbestos ferrolite	2.5	20
Compressed asbestos fiber	3	25
Camprofile	2.25	27.5
Thermoculite	3.2	34.5
Unreinforced graphite laminated Klingersil	4.5	45
Reinforced graphite laminated Klingersil	2-5	50
Flat metallic (Al) double gasketed graphite filled	3.25	38
Flat metallic (SS) double gasketed graphite filled	3.75	62
Spiral wound metal (CS) mineral fiber filled	2.5	69
Spiral wound metal (SS) mineral fiber filled	3	69
Solid flat metal or O ring, soft Aluminum	4	60
Solid flat metal O ring, soft carbon steel	5.5	120
Solid flat metal O ring, soft stainless steel	6.5	175
Kammprofile	2	18

13.2.4 Philosophy of the Selection of Flange Parameters

The following guidelines can be used to select parameters based on the above basics and code rules for given design data.

1. Select the gasket type and width based on pressure, temperature, and availability in the market.
2. Select the bolt material depending on the flange/pipe material. Assume the preliminary lowest size limiting quantity for ease of maintenance.
3. Calculate flange dimensions C, G_O, and G using R and E values for the bolt size required to calculate the bolt forces in the operating and seating condition (W_0, W_a) and required thickness (t) of flange including limits of P_C as illustrated in examples in Tables 13.3 and 13.4.
4. Calculate the number (n) of bolts required to withstand the above forces, increase to nearest integer multiple of four, and calculate pitch of bolts (P_C).
5. If P_C is less (normally due to high pressure), increase the bolt size which will increase both but Pc at a higher rate than the min limit as the area of bolt increases with square of size increase. If the increase up to max available size (100 mm) is not adequate, increase BCD.
6. If P_C is more, increase the number of bolts or reduce the size

13.3 FLANGE ANALYSIS

For a given diameter of pipe, pressure, temperature, and material, first select the bolt size and compute dimensions of the flange as explained in 13.2. The flange is then analyzed in both operating and seating conditions. Refer Figure 13.2 and Tables 13.3 and 13.4 for notation and forces.

13.3.1 Seating Condition

The gasket (except full faced) when tightened before operating develops a reaction force due to its elasticity and induces moment M_a at the gasket reaction circle (diameter *G*) due to bolt load (*W*) which depends on the operator. Normally maximum torque for tightening bolts is given in operating instructions of equipment. The minimum value of *W* is equal to maximum bolt load Wma to compress gasket to obtain pressure *y* or Wmo in operation to overcome the forces $H + Hp$ due to pressure and gasket loads, and the maximum value is equal to the bolt stress reaching its allowed stress. The optimum value is the average of both minimum and maximum considered.

Also the flange will rotate at gasket reaction diameter *G* due to the moment at *G* due to force *W* by tightening bolts. Due to this rotation, the circle at point *G* which without tightening is the midpoint of the gasket width will shift toward C, and gasket pressure will be triangular and leak in operation. Therefore, rotation shall be kept under limit. However, *G* is taken as the mid-point of the gasket if the width is <12.5 mm; otherwise *G* is equal to gasket OD minus twice effective gasket width. The loading diagram is force at the bolt circle and support at gasket reaction circle *G* on a circular ring. Forces and details of loose and integral flanges are shown in Figure 13.3.

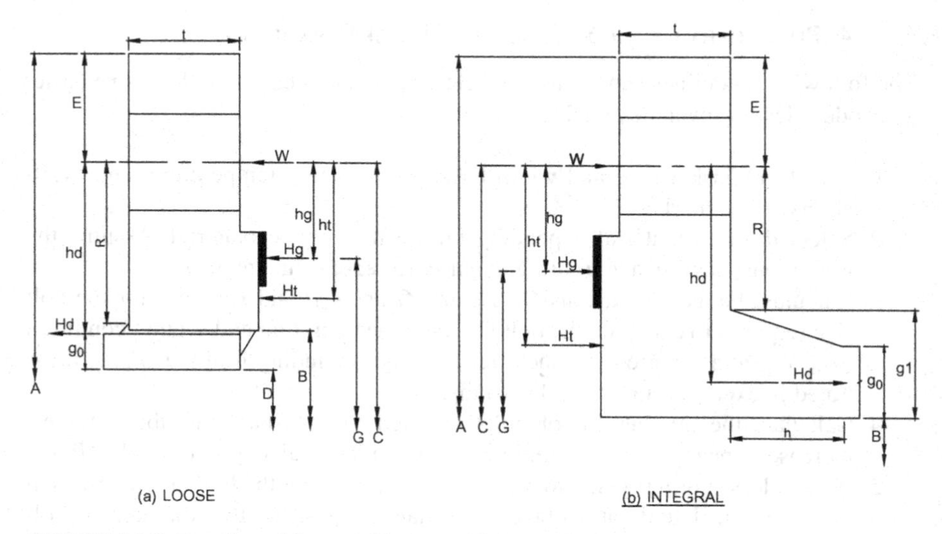

FIGURE 13.2 Details and forces of loose and integral flanges.

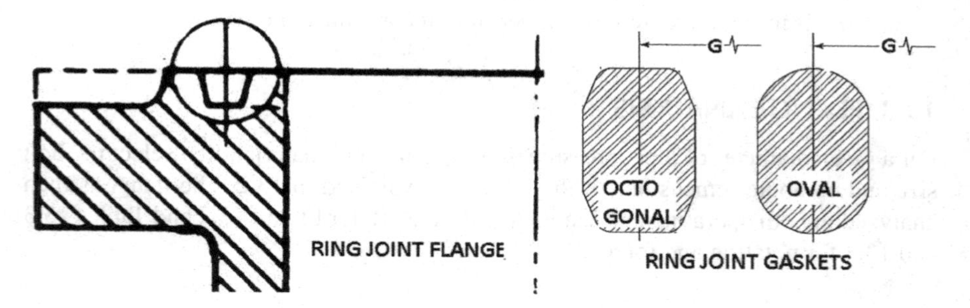

FIGURE 13.3 O ring type joints.

13.3.2 Operating Condition

- F_d = Due to pressure on the internal area of the pipe acting on the pipe wall thickness or at B.
- F_g = Due to the gasket reaction acting at about midpoint (gasket reaction diameter) of gasket width. F_g shall be m times the gasket reaction due to y on the area of the gasket contact surface with effective width b (including pass partisan gasket if any) to keep sufficient force for leak proof.
- F_t = Due to the pressure on the annular portion between the gasket reaction and pipe inside diameter acting at the midpoint of annular portion.

All these loads induce moment M_o at the bolt circle diameter and rotation at G. Point G will shift outwards as explained above and leak. Hence, rotation shall be limited to keep the gasket pressure almost uniform. Rigidity of the flange is calculated apart from longitudinal stress in hub and radial and tangential stresses in the flange.

The rigidity equation is based on limiting rotation to about 1/3 to ½°. The code gives an equation for rigidity in both conditions and for both loose and integral flanges.

Flange stresses: radial, tangential, and longitudinal in hub in operating and seating conditions are derived by integration using the loading diagram shown in Figure 13.1 as (constant x M/B)/Z. where

Z = the section modulus per unit width ($Z = t^2/6$) for the flange and ($g_1{}^2/6$) for the hub

$M = M_O$ for the operating condition and M_a for the seating condition.

Constants depend on $k = A/B$ and hub dimensions and are given in Ref. 2 and code.

13.4 COMPUTATION OF STRESSES IN FLANGES

13.4.1 Loose and Integral Flange

Integral flanges require forging and are generally selected for high pressures. Loose flanges can be cut from plate for low pressures. For higher pressures, it is advised to roll a plate and weld to form a ring. Rolling will give the required metallurgical structure similar to forging. The thickness required for the integral flange is less than that for the loose flange. Because the thickness of the pipe reduces the flange thickness, the integral type is selected for thicker pipes. Hub dimensions have large influence on constants of proportions thereby in flange thickness. An increase in the hub length and reducing taper angle will reduce the flange thickness as the hub will offer more resistance. See Figure 13.2 with loading for the loose flange with a raised face and integral flange.

A loose flange without a hub is a flat circular ring assumed as isolated with forces shown in Figure 13.2a and reaction force and moments at G and C for seating and operating conditions. Basic analysis is covered in Chapter 10. Tangential and radial stresses are induced due to forces. Stresses can be calculated with certain assumptions from equations given in Table 24 of Ref. 3. By integration, stresses can be derived for all the forces shown in Figure 13.2a. It is found that max moment is tangential and at bolt circle (C) for the operating condition, and tangential at gasket reaction dia (G) for the seating condition. Ref. 2 gives integration analysis, and the tangential stress is given as

$$\mathrm{YM}/\left(\mathrm{B\,t^2}\right)$$

where M is Mo and Ma for operating and seating conditions

Y is constant depending on K = A/B and code gives the equation for Y as

$$Y = \left[0.66845 + \frac{5.7169K^2 \log K}{K^2 - 1}\right]/(K-1) \tag{13.1}$$

Table 13.3 illustrates the detailed calculations of stresses and rigidity check for the loose flange.

TABLE 13.3
Calculation of Stresses in the Loose Flange without Hub Figure 13.2

Data: units: N, mm, MPa; pressure = 1, temp. = 225°C, mat. SA-226,

Bolt data

a	24	Nominal diameter of bolt
n	72	Number of bolts
A_O	312.75	Min root area of bolt

Gasket data: type - flat metallic graphite filled

m	3.75	
y	62	Gasket seating load
N	19	Gasket contact width
b	7.71	Eff. Gasket seating width = min(N/2,2.5$\sqrt{[N/2]}$

Bolt spacing

R	36.51	ID to BCD, (C-B)/2
E	28.58	BCD to OD, (A-C)/2
P_C	91.8	Bolt pitch min = 2a + 6 = 54, max = 2a + 6t(m + 0.5) = 316

S-allowed stresses, E-Young's modulus

operating.o		Ambient.a, f-flange, b-bolt
S_f	137	138
S_b	172	172
E	185000	202000

Flange data

D	2000	Pipe inside diameter
T	10	Pipe thickness
t	190	Flange minimum thick
B	2023	Flange ID = D + 2T + 3
C	2106	BCD = B + 2R +10
A	2163	Flange OD = C + 2E
R_f	2070	Raised face OD = C – a – 12
G_O	2067	gasket OD = R_f – 3

Bolt load, bolt area, Forces and moments

G	2052	Gasket reaction diameter = Go – 2b
H_P	372481	Contact load on gasket surface = π G 2b P m
H	3305762	Total. hydrostatic end force = π/4 G^2 P
Wmo	3079178	minimum operating bolt load = H + Hp
Wma	3678243	Minimum gasket seating bolt load = π G b y
A_m	21385	Total required area of bolts, max(W_{ma}/S_b, W_{mO}/S_a)
A_b	22518	Total provided area of bolts = A_O n
W	3873071	flange design bolt load, seating, 0.5(A_b + A_m) S_a, for max A_b S_a
H_g	372481	gasket seating load, W_{ma} – H
H_d	3214272	hydro static end force on area inside of flange, π/4 B^2 P
H_t	91490	Pressure force on flange face, H – H_d
h_g	27.2	Radial distance from BCD to the circle on which load H_g act = (C – G)/2
h_d	41.5	Radial distance from BCD to the circle on which load H_d act = (C – B)/2
h_t	34.4	Radial distance from BCD to the circle on which load H_t act = {h_d + h_g)/2
M_O	146 KNm	Total moment in operation = $M_g + M_d + M_t = H_g h_g + H_d h_d + H_t h_t$
Ma	105 KNm	Total moment for gasket seating, W h_g
K	1.0692	K = A/B Flange constant Y 28.83 [0.668 + 5.72{K^2LOG(K)} – 1]/(K – 1)

	Operating	Seating	Stresses induced in loose flange w/o hub
St	57.9	41.6	Tangential flange = Y(M_O, Ma)/(B t^2), allowed = S_f
B_S	163	136	Bolt stress = {W_{mo}, W_{ma}}/A_b, allowed = {S_{bO}, S_{ba}}
J	0.945	0.679	Rigidity = 109.4(M_O, Ma)/[E t^3 0.2LN(K)] < 1

Analysis of the integral flange is more complicated due to the discontinuity at pipe to hub, hub to flange ring, and resistance of pipe size and thickness. However, the compilation procedure is the same as that for the loose flange except the constants depend on B, g_o, and g_1 in addition to K. Furthermore, max stress may be radial or tangential in the flange, and longitudinal stress in the hub also is to be checked. Ref. 2 gives integration analysis and derives similar equations to those for the loose flange. Unlike the loose flange, stresses in the integral flange include longitudinal stress in the hub and radial stress in the flange apart from tangential stress in the flange and are given as proportional to M/(B t^2) for the flange and M/(B g_1^2) for the hub. Constants are given in the code and Ref. 2. Table 13.4 illustrates the detailed calculations of stresses

TABLE 13.4

Calculation of Stresses in the Integral Flange and Loose Flange with hub Figure 13.2

Data: units: N, mm, MPa UOS; pressure = 1, temp. = 225°C, mat. SA-226, flange thickness = 150

Bolt data

a	24	Nominal diameter of bolt
n	72	Number of bolts
A_O	312.75	Min root area of bolt

Gasket data: type - flat metallic graphite filled

m	3.75	
y	62	Gasket seating load
N	19	Gasket contact width
b	7.71	Eff. Gasket seating width = min(N/2,2.5$\sqrt{[N/2]}$)

Bolt spacing

R	36.51	ID to BCD, (C-B)/2
E	28.58	BCD to OD, (A-C)/2
P_C	91.8	Bolt pitch min = 2a + 6 = 54, max = 2a + 6t(m + 0.5) = 316

S-allowed stresses, E-Young's modulus

Operating.o Ambient.a, f-flange, b-bolt

S_f	137	138
S_b	172	172
E	185000,	202000

Flange data

B	2000	Flange inside diameter
g_O	10	Hub thickness at small end
g_1	15	Hub thickness at large end
h	15	Hub length ≥ 1.5 g_O

C	2106	BCD = B + 2(g_1+R)
A	2163	Flange OD = C + 2E
R_f	2070	Raised face OD = C – a – 9
G_O	2067	Gasket OD = R_f – 3

Bolt load, bolt area, Forces and moments

G	2052	Gasket reaction diameter = Go – 2b
H_P	372481	Contact load on gasket surface = π G 2b P m
H	3305762	Total. hydrostatic end force = π/4 G^2 P
Wmo	3079178	Minimum operating bolt load = H + Hp
Wma	3678243	Minimum gasket seating bolt load = π G b y
A_m	21385	Total required area of bolts, max(W_{ma}/S_b, W_{mO}/S_a)
A_b	22518	Total provided area of bolts = π/4(root dia)^2 n
W	4948956	Flange design bolt load, seating, 0.5(A_b + A_m) S_a, for max A_b S_a
H_g	372481	Gasket seating load, W_{ma} – H
H_d	3141600	Hydro static end force on area inside of flange, π/4 B^2 P
H_t	164162	Pressure force on flange face, H – H_d
h_g	27.2	Radial distance from BCD to the circle on which load H_g act = (C – G)/2
h_d	45.5	Radial distance from BCD to the circle on which load H_d act = (C – B – g_1)/2
h_t	40.1	Radial distance from BCD to the circle on which load H_t act = [(C – B)/2 + h_g]/2
M_O	160 KNm	Total moment in operation = M_g + M_d + M_t = H_g h_g + H_d h_d + H_t h_t
Ma	105 KNm	Total moment for gasket seating, W h_g

Flange factors for integral or loose with hub

X_g	1.5	g_1/g_o
h_O	141.4	$\sqrt{(B*g_o)}$
X_h	0.106	h/h_o
K	1.0815	K = A/B
Z	12.79	$(K^2 + 1)/(K^2 – 1)$
L	5.181	$(t e + 1)/T + t^3/d$
e	0.00639	F/ho in 1/mm

T	1.878	[K^2{1 + 8.55LOG(K)} – 1]/[(1.05 + 1.94K^2)(K – 1)]
F	0.904	Figure 2.7.2 for integral, 2.7.4 for loose of code
U	27.01	[K^2{1 + 8.55LOG(K)} – 1]/[1.36(K^2 – 1)(K – 1)]
V	0.468	Figure 2.7.3 for integral, 2.7.5 for loose of code
d	8.16E5	U h_O g_O^2/V
Y	24.66	[0.668 + 5.72K^2 LOG(K)/(K^2 – 1)]/(K – 1)
f	1.824	Figure 2.7.6 of code for integral, 1 for loose flange

Stresses induced

	Operating	**Seating condition**	
S_h	124.9	82.44	Longitudinal hub = (f/L){M_O, M_a}/(B g_1^2), all = 1.5S_f
S_r	1.558	1.028	Radial flange = {M_O, M_a}/(B t^2) (1.33t e +1)/L, all = S_f
S_t	67.57	44.59	Tangential flange = {M_O, M_a}/(B t^2)Y – Z Sr, all = S_f
S_a	96.24	63.52	Average flange = [S_h + max(S_r, S_t)]/2, all = S_f
B_S	163.3	137	Bolt stress = {W_{m1}, W_{m2}}/A_b, all = {S_b, S_a}
J	0.959	0.58	Flange.rigidity.index = 52.14V(M_O, M_a)/[L(E_O, E_a)g_O^2 k h_O] k = 0.3 for integral, 0.2 for loose

and rigidity check for the integral flange using the same data of example of Table 13.3 for the benefit of comparison. Except B, H_d, H_t, h_d, h_t, and M_o, for flange factors and stresses, all data and calculations are the same. Using the integral flange instead of loose flange, the thickness required for the flange is reduced from 190 to 150 mm.

Analysis of the loose flange with the hub is the same as the integral flange except flange factors *F*, *V*, and *f*. The values of *F* and *V* increase, and that of *f* will reduce. The effect of *F* will increase stresses and reduce rigidity, that of *V* reduces stresses and increases rigidity, and that of *f* decreases stresses. Generally, the effect is negative as the hub to pipe joint is not integral. The constants *F*, *V*, and *f* can be read from graphs in the code.

13.4.2 Full Faced Flange

Large raised face flanges with low pressures are normally subjected to higher seating stresses than operating. The flat faced flange with the gasket within bolts is the same as 13.4.1 as long as both flanges will not touch at end due to bolt tightening. However with a full gasket, moment in the seating condition is negligible although bolt load is higher due to a large contact surface. Therefore, full faced gaskets are used for large flanges and low pressures with gaskets made of non-metallic material and with low compressive strength. Analysis in the operating condition is the same as the example in Table 13.3. Additionally, the gasket portion outside the bolt circle will exert balancing reaction H_r at a distance h_r from G_o, opposing M_o and is given by

$$H_r = M_o / h_r$$

where $h_r = (Go - C + d_h)/4$ and d_h is the bolt hole diameter

Operating bolt load W_O will be higher by H_r and the bolt required area in the operating condition will go up. Therefore, apart from the above (W_O and Ma), the analysis is the same as that for loose and integral flanges given in Table 13.3 and 13.4.

13.4.3 Ring-Type Joints

Ring-type joints are used for high pressures with soft iron or steel oval or octagon ring gaskets (R type). Gaskets will have a low width and higher height. The grooves in both mating flanges in which the ring gasket seats will be self-confining as required for high pressures. The analysis is the same as the integral flange except m and y values, which are much higher, m = 6.5 and y = 179 MPa.

13.4.4 Lap Flange

A lap flange is a flat plate circular ring welded to the companion flange at the outer edge. Two loads are acting on it. First, (F_1) force per unit circumference at its inner edge due to pressure in the pipe and second, pressure load on the flange surface (area of ring between OD and ID of flange) as shown in Figure 13.4. The calculation of moments and stresses is based on flat plate theory for the circular ring given in 10.4.2.

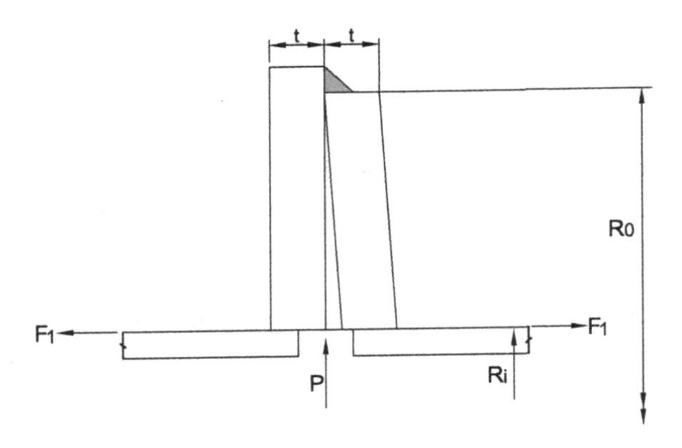

FIGURE 13.4 Lap flange.

Lap flanges are used for large sizes and low pressures to reduce the cost and flexibility of alignment and re-welding. The flange width shall be taken minimum as possible (50-75mm) to reduce thickness. Weld throat shall be adequate to resist tensile load due to pressure over area to radius Ro. Companion flange OD shall be more than Ro to accommodate fillet weld. Both flanges can be welded inside if access is available.

13.4.5 Blind Flange

Blind flange analysis is the same as that for the loose flange except the central portion is subjected to pressure the effect of which is added stress equation. The stresses are derived below:

Seating condition: As moment Ma is at gasket reaction diameter G, section modulus at G is $\pi G\, t^2/6$, and tangential stress is given by

$$M_a/z = 6/\pi\, M_a/(G\,t^2) = 1.91 M_a/\left(G\,t^2\right) \tag{13.2}$$

Operating condition: Apart from the above equation with Mo replacing Ma, additional stress is induced due to the pressure acting on blind flange up to diameter G. Tangential stress can be calculated from flat plate theory assuming the boundary condition. However, the boundary is neither fixed nor SS, and stress is approximately equal to
$0.3P(G/t)^2$. Therefore, adding both tangential stresses is

$$1.91 M_a/\left(G\,t^2\right) + 0.33P(G/t)^2 \tag{13.3}$$

13.4.6 Rectangular Flange

Rectangular flanges are used normally for low pressures and slip on type similar to the loose circular flange without a hub and with or without the raised face. Their cross-section is the same as that of the circular flange. Normally, non-metallic gaskets with low compressive strength (y value) are used. Analysis is based on flat plate theory. The pressure effect (force) is in two directions, normal to flange face and tangential.

13.4.6.1 Analysis in the Normal Direction

The pressure effect in normal to flange face is the same as the axial direction of circular flanges, and the same notation can be used replacing diameters with suffixes a and b. These suffixes may be omitted for dimensions from inside to outside of the flange as they are the same in four sides. Forces and other details are shown in Figure 13.5. Furthermore, the following assumptions are made which are conservative and valid for large ducts.

- Gasket nominal width can be full face or any width up to bolt without raised face or up to raised face. The max gasket width other than full face from B to bolt gives a theoretical gasket reaction at a distance (C – d/2 – B)/2 from *C*. Practically it is less, and the effective gasket width is very less depending on the type of gasket. Ref. 4 prescribes 20mm as the effective width *b* and gasket force considered as acts at distance h_g from *C*, $h_g = f_g(C - B)/4$ omitting the effect of bolt size, f_g (≥ 1) is the geometrical correction factor depending on the max gasket width for partial gaskets. For gaskets beyond the bolt circle (full face), f_g is not applicable and taken as one.

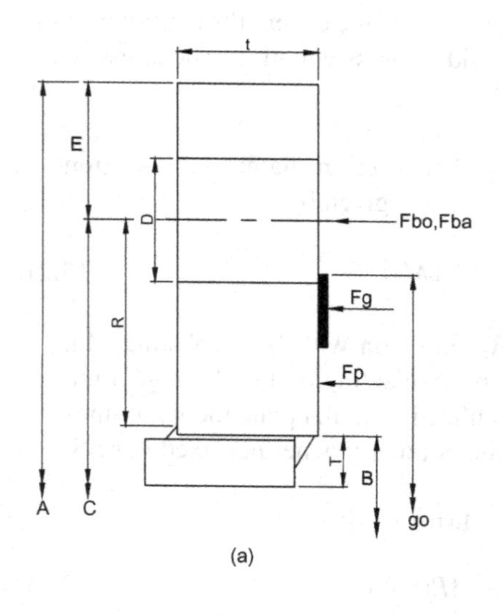

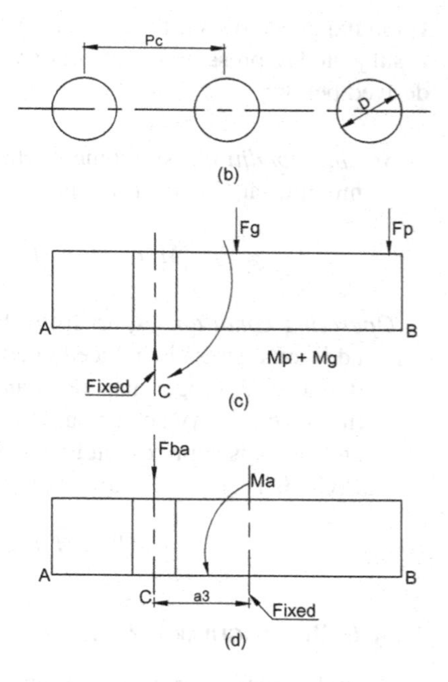

FIGURE 13.5 Rectangular flange.

- The effect of pressure force is the same as that of the circular flange; however, it can be considered as single force with pressure up to the bolt center and acts at about the mean thickness of duct.
- Beam theory can be used because the flange width is very small compared to the duct dimension.
- Stresses are the same in all ligaments so that analysis is limited to one ligament or for one bolt provided the pitch is constant.
- Bolt size (d) is advised min 20 mm, and pitch shall be optimum for the quantity of bolts and flange thickness.

The forces and moments in operating and seating conditions are derived below with the above considerations.

F_p = pressure force = $P\ C_a\ C_b$ acts at a distance $[h_p = f_p(C - B + T)/2]$ from C

$f_p = 0.9 + (C - B)/2 - a_3)/a$ is the geometrical correction factor and = 1 if the gasket is beyond bolt circle (full face)

F_g = gasket force = $y\ [G_{oa}\ G_{ob} - (G_{oa} - 40)(G_{ob} - 40]$ acts at $[h_g = fg(C - B)/8]$ from C

F_{bo} = total bolt load operating = Fp + Fg

F_{ba} = max bolt force in the seating condition = $N\ A_o\ S_{ba}$ acts at C for raised face or gasket within bolt, not applicable for the full faced gasket

where

T = thickness of duct + gap between the flange and duct

N = number of bolts

A_o = net area of bolt

S_{ba} = bolt allowed stress at ambient temperature

Considering the rectangular flange as a flat beam of width equal to pitch of bolts in both operating and seating conditions, stress analysis is simple with insignificant error.

M_a = Moment per pitch due to $F_{ba} = F_{ba}\ a_3/N$

$hg = fg(C - B)/8$

Bending stress in the seating condition = M_a/Z

Min number of bolts required = $F_{bo}/(S_{bo}\ A_o)$

M_P = moment per pitch due to $F_p = F_p\ h_P/N$

f_p = geometric correction factor = if$\{h_P > 1.2e_2, 0.9 + (h_P - e_2)/e_2, 1\}$

M_g = moment per pitch due to $F_g = F_g\ h_g/N$

f_g = geometric correction factor = if$\{(C - B)/2 > a_3, 0.9 + (C - B)/2 - a_3)/a_3, 1\}$

Bending stress in operating due to $M_P + M_g = (M_P + M_g)/Z$

where

$e_2 = (A - B + T/2)/4$
$a_3 = (G_o - B)/4$
$Z = (P_C - D)t^2/6$ for operating and $P_c\ t^2/6$ for seating
P_C = pitch of bolts
D = diameter of bolt holes

Equations for geometric correction factors f_P and f_g are taken from Ref. 4.

Example 13.1: Calculate flange stresses for design data: pressure 0.1 MPa, temperature 250°C, rectangular duct 2000 × 1500 inside, 10 thick, SA-516 Gr.70, allowed stress 138 MPa in both conditions, 122 × M24 bolts SA-193 B7 allowed stress 172 MPa at both conditions, and soft gasket with 20MPa compressive strength. Units are in *mm* and *N*. Symbols as shown in Figure 13.5

- Bolt parameters R = 37, E = 27, and $A_O = 313\ m^2$
- B = 2000, 1500, C = B + 2R = 2074, 1574, A = C + 2E = 2132, 1632, N 122, Pc = 60, a-bolt nominal dia = 24, gasket OD = G = C – a = 2050, and T = duct thickness + gap =12
- Number of bolts required = $2(Ca + C_b)/\ P_C = 121.6$
- $F_p = P\ Ca\ C_b = 0.1*2074*1574 = 326448$
- $F_g = y[G_a\ G_b - (G_a - 40)(G_b - 40)] = 2848000$
- $F_b = F_p + F_g = 3174448$
- $a_3 = (G - B)/4 = 12.5$
- $h = (C - B)/2 = 37$
- If [$h > a_3$, $fg = 0.9 + (h - a_3)/a_3 = 2.86$, (else fg = 1)
- $hg = fg\ h/4 = 26.455$
- $Mg = Fg\ hg/N = 617572$
- $e_2 = (A - B + T/2)/4 = 34.5$
- If{$h + T/2 > 1.2e_2$, $f_P = 0.9 +(h + T/2 - e_2)/e_2 = 1.146$, (else $f_P = 1$)
- $h_P = fp(h + T/2) = 49.3$
- $M_P = F_P\ h_P/N = 131901$
- $Z = (P_C - d)\ t^2/6 = 5632$
- $\sigma_f = (Mg + M_P)/Z = 133 < 138$
- $F_{ba} = N\ A_o\ S_{ba} = 92*313*172 = 53836$

$\sigma_S = Ma/Z_S = F_{ba}\ a_3/(p_C\ t^2/6) = 672950/10240 = 65.7 < 138$

13.4.6.2 Analysis in the Tangential Direction

Pressure not only acts on the flange face, but the pressure force acting on the portion of connecting duct partially transfers to the flange and depends on the arrangement of duct with or without stiffeners. Refer section 10.5 for the calculation guidelines of the above force. Each side of the flange with an effective width of duct can be considered as a fixed beam (T shaped) under the loading explained above.

13.4.7 Standard Flanges

A standard flange is predesigned, and several countries have their own standards for flanges; however, the widely used one is the American Standard. ASME B16.5 provides all the required details for sizes up to NPS 24" and B16.47 for sizes from NPS 26–60".

Flanges are classified by size and class. The class is number, and its value is approximately equal to pressure in psi. The standardization (design) is based on all the other related parts like gasket, bolts etc., and materials are used as given in the standard.

The classes in B16.5 are: 150, 300, 400, 600, 900, and 1500 up to 24"; 2500 up to 12".

The class is selected from rating tables for the material, pressure, and temperature available in standard. Pressure, temperature, and class can be calculated by the following equations

For 150 class

Ceiling pressure = 21.41–0.03724T bar

Ceiling temp = (21.41–P)/0.0372°C

For 300 class and above

Class = 8750P/S,

where

S = stress at temperature in bar, P = pressure in bar, T = temperature °C

If 2500 class is not adequate, increase the bolt size or number of bolts.

The classes in B16.47 are 75, 150, 300, 400, 600, and 900 A & B series.

In A series, the bolt size is higher and OD of flange is more over B series, B series are compact. Pressure, temperature in °F, and class can be calculated using the following equations

Class 75

Ceiling pressure = 160–0.15T

Ceiling temp °F = (21.41-P in psi)/0.03724

Class 150

Ceiling pressure = 320–0.3T
Ceiling temp °F = (21.41-P in psi)/0.03724

Class >150
Class = 8750 P/S

13.4.8 Other Flanges

Noncircular flange with a circular bore: The outside diameter *A* for a noncircular flange with a circular bore shall be taken as the diameter of the largest circle, concentric with the bore, inscribed entirely within the outside edges of the flange. Bolt loads and moments, as well as stresses, are then calculated as per circular flanges, using a bolt circle drawn through the centers of the outermost bolt holes. For external pressure the same as integral except

$$M_O = H_d(h_d - h_g) + H_t(h_t - h_g)$$

Reverse flange: Reverse flanges are reducing with a larger flange bolt circle inside and less than outside diameter of the shell. They are used in special applications and under external pressure. The analysis is covered in code.

High pressure flanges: These are used for very high pressures for equipments such as feed water heaters, and pressure will give seating force.

REFERENCES

1 Code ASME S VIII D 1, 2019.
2 *Process equipment design*, L. E. Brownell & E. H. Young.
3 *Formulas for stress and strain*, Raymond J. Roark & Warren C. Young, 5th edition.
4 Heat Exchanger Institute standard.

14 Vibration

14.1 GENERAL

Vibration in pressure vessels is mainly in tubes, tube banks, rectangular enclosures to tube banks, and rectangular ducts. These vibrations are flow induced, either inside flow of fluids or outside. The vibration subject starts with natural frequency (f_n). Natural frequency of any element is the measure of its resistance to vibration and is given by Eq. 14.1. Minimum natural frequency advised is about 2.5 to 3 to prevent vibration.

$$f_n = \frac{\sqrt{g/y}}{2\pi} \tag{14.1}$$

where

y = deflection of the element simply under its self-weight

g = gravitational constant = 9.81 m/s^2 or 32 fps^2

Depending on the arrangement of elements in relation to connecting elements in equipment, other factors such as fluid will contribute to load as well as for damping.

For the purpose of analytical vibration analysis, elements are mainly two types

- Beam element
- Plate element

Others such as the three-dimensional element and whole equipment are complicated to analyze analytically and require CFD (computational fluid dynamics) & FEA (finite element analysis).

For the purpose of f_n, the *element* is a body or part of any equipment, which can be isolated from rest of the equipment with well-defined boundaries whose f_n can be calculated by mathematical equations. Part of any piping system between two adjacent supports, a tube of shell and tube exchanger between tube sheets, a stiffened flat plate or part of it enclosed by stiffeners, or any part of equipment with clear boundary conditions which can be isolated to arrive at its free body diagram are examples of elements.

Types of vibrations are vortex shedding, buffeting, and acoustic. Fluid flowing past any element induces frequency which coincides with natural frequency of any mode, the amplitude multiplies, and the element vibrates.

DOI: 10.1201/9781003091806-14

14.2 DAMAGING EFFECTS OF VIBRATION

The structural effect of vibration is basically fatigue failure. Damaging effects are:

1. Vibration of element due to large amplitude has impact on tubes against each other or the end tube with the vessel wall.
2. Wear of the tube and tube holes in baffles and tube sheets.
3. Vibrating element fails by fatigue.
4. Process performance is affected.

14.3 NATURAL FREQUENCY

The lowest frequency of an element is called fundamental or the 1st mode and involves deflection of elements at its middle point. The 2nd mode frequency is higher and involves deflection at two points and so on. For equipment with several elements, the most flexible element will have lowest f_n and is more prone to vibration. The 2nd mode frequency can be in the same element at other points or in other elements. f_n depends on deflection y which can be calculated from the loading diagram. Apart from self-weight, other factors such as weight of fluid inside, weight fluid displaced by element, and hydrodynamic mass contribute wherever applicable. If the element is in tension, its effect is added to Eq. 14.1, by multiplying factor >1 (A). Outer tube bend f_n is lower than inner. These effects and computation of *fn* of some normal elements of pressure vessels are given in the following sections.

14.3.1 TUBES OF THE SHELL AND TUBE HEAT EXCHANGER

The derivation of f_n is based on structural basics and example 14.1 illustrates the procedure.

Example 14.1: Calculate natural frequency given the following data:

p = Square pitch = 76.2 mm

d = Tube outside diameter = 50.8 mm

t = Tube thickness = 4 mm

d_i = Tube inside diameter = 42.8 mm

L = Tube length = 6 m

ρ_s = Shell side fluid (water) density = 800 kg/m^3

ρ_t = Tube side fluid (gas) density = 0.56 kg/m^3

ρ = Tube material density = 7850 kg/m^3

F = 10000 kgf, tension in the tube due to the tube staying as the tube sheet against deflecting under shell side pressure = (tube supporting cross-sectional area x shell side pressure)

E = 19000 kgf/mm^2

I = Moment of inertia of tube = 162106 mm^4

Computation of deflection (y) of tube in Eq. 14.1

W_t = Weight of the tube = $0.25\ \pi d_i^2\ \rho$ = 0.00461 kg/mm

W_i = Weight of fluid in the tube = $0.25\ \pi d_i^2\ \rho_t$ = 8.1E-7 kg/mm (neglected)

W_d = Weight of fluid displaced by tube/unit length = $0.25\ \pi d^2\ \rho_s$ = 0.00162 kg/mm

C_m = 1.275 for P/d = 1.5, added mass coefficient depends on p/d (P = longitudinal pitch) and given in table below (ref: TEMA)

P_L /d, P_L = long pitch 1.5, 1.4, 1.3, 1.25, 1.2

Staggered 1.28, 1.36, 1.53, 1.7, 1.97

In line 1.275, 1.51, 1.425, 1.51, 1.6

H = Hydrodynamic mass = $W_d\ C_m$ = 0.00206, H increases the apparent weight of the vibrating body due to displacement: 1) motion of the body, 2) proximity of the tubes within the bundle, and 3) relative location of the shell wall

W_o = total effective weight per unit length = $W_t + W_i + H$ = 0.00667 kg/mm

$y = 5W_oL^4/(384EI)$ from beam formulas for simply supported = 37 mm

$F_c = (k/L)^2EI$ = 843.5 kg, where k depends on the boundary condition = 3.14 (π for SS, 4.49 for one end fixed & other SS, 2π for fixed refer TEMA). Tube ends are assumed as simply supported

A = factor for effect of tube tension $\sqrt{1+F/F_C}$ = 3.585, 1 if no tension

$f_n = \dfrac{A\sqrt{g/y}}{2\pi}$ = 3.585/6.28*$\sqrt{}$(9810/37) = 9.34 cycles/sec

Because natural frequency is more than three, the tube may not vibrate.

14.3.2 Tubes of the Tube Bank of Heat Exchanger

The derivation of f_n is similar to example 14.1. Example 14.2 illustrates the procedure.

Example 14.2: Same data as example 14.1, except fluid in the tube side is water and that in the shell side is gas, and no tension in tube.

The following values will change:

A = 1 as F = 0, because no tension

$W_i = 0.25\ \pi d_i^2\ \rho_t$ = 0.00115 kg/mm

$W_d = 0.25\ \pi d^2\ \rho_s$ =1.13E-6 kg/mm

$H = W_dC_m$ = 1.44E-6

$W_o = W_t+W_i+H$ = 0.00461+0.00115 = 0.00576 kg/mm

$y = 5W_oL^4/384EI$ = 5*0.00576*6000^4/(384*19000*162106) = 31.5 mm

fn = A$\sqrt{}$[g/y]/2π = 1/6.28*$\sqrt{}$(9810/31.5) = 2.81 cycles/sec

Since the natural frequency is less than three, the tube may be prone to vibration.

14.4 VIBRATION IN FLAT PLATES

The notation is the same as in example 14.1

Normally, inside of the plate is fluid (mostly gas) in pressure vessels; the other side is open to atmosphere. Vibration may be due to inside fluid flow or outside wind air or induced due to vibration of internal elements such as tube bundle or combination. For calculating y, only its self-weight (W_t) including stiffeners if provided is effective. W_i is not applicable, H is negligible, and therefore $W_o = W_t = \rho \, x \, t$, where t = thickness of the plate

Deflection y can be calculated as per applicable equation for load W_o and boundary conditions derived from plate theory (Chapter 10), and f_n is calculated from Eq. 14.1 $= \frac{\sqrt{g/y}}{2\pi}$

Normally, the rectangular duct or chamber or vessel which includes flat elements is considered for vibration and for calculating f_n. It is difficult to isolate or arrive at the free body diagram of rectangular plate elements from such equipment to calculate f_n. However, flat elements with known boundary conditions with attachments like stiffeners, f_n can be calculated by evaluating thickness (te) of equivalent flat plate for stiffener by equating moment of inertia of both. $te^3 = 12Is/p$, Is = MI of stiffener, p = pitch of stiffeners attached over the entire plate. If attached in both directions, p is replaced by p/2.

14.5 TALL VESSELS LIKE CHIMNEYS

For equipments such as chimneys, f_n can be calculated by arriving at equivalent deflection y by the summation procedure.

$$y = \Sigma(m\, y'^2)/\Sigma(m\, y')$$

where

m = weight of the element (segment)
y′ = deflection at CG of each element due to m at CG

14.6 FLOW-INDUCED FREQUENCIES AND VIBRATION

Fluid dynamics and engineering standards specify normal working velocities of fluids. The basis of which is economics of capital and operational costs. Pressure parts except tubes & pipes generally have *fn* much above flow induced frequencies. Vibration problems are limited to low pressure parts or high fluid velocities for certain equipment's and parts with low f_n. Constraints in providing more supports or restraints than required for pressure and static design, and deviating from normal plant layout due to space, height, road restrictions, etc. will result in abnormal flow and non-uniform velocities, which will induce high frequencies and be prone to vibration. Flow-induced frequencies are the following three types.

- Vortex frequency f_v
- Buffeting frequency f_b
- Acoustic frequency f_a

When vortex or buffeting frequency coincides natural frequency of element, it will vibrate. Practically, vibration will be felt when f_V or f_b is in the range from $0.8f_n$ to $1.2f_n$, peak at f_n. When f_v *or* f_b crosses $1.2f_n$, vibration will not be felt until frequency increases nearer to f_n of second mode. Flow-induced vibration in tubes will cause vibration in enclosed casing if its natural frequency is close. Computation of flow-induced frequencies for tubes of the tube bundle is given in the following sections.

14.6.1 Vortex Frequency

Vortex shedding is a phenomenon, when the fluid flows across a structural member such as tube, vortices are shed alternately from one side to the other, as alternating low-pressure zones are generated on the downstream of the structure, giving rise to a fluctuating force acting at right angles to the flow direction. The *Strouhal Number* (S) is a measure of the ratio of the inertial forces due to the unsteadiness of the flow or local acceleration to the inertial forces due to changes in velocity from one point to another in the flow field. S depends on the pattern of tubes (inline or staggered or otherwise) and cross or longitudinal pitch. S for tubes in tube bank can be read from Table 14.1 (ref: TEMA). Vortices are formed when fluid flows in any conduit whenever gross discontinuities are encountered in its flow path. If the induced frequency is close to the f_n of the conduit wall local to vortices, the wall vibrates. The elements of such conduits shall be stiffened, and the parts transferring the load to the support shall be stiff enough to obtain large f_n. Furthermore, discontinuities shall be minimized using the following precautions:

1. Increase or decrease of the cross section where required shall be gradual and divergence or convergence shall not be more than 30°. If it is not possible, either

TABLE 14.1
Strouhal's Number

Xt	xl	Inline	Staggered
1.3	1.25	0.42	0.4
	1.5	0.25	*
	2	0.15	0.21
	2.5	0.12	0.15
	3	0.05	0.1
1.5	1.25	0.46	0.6
	1.5	0.3	*
	2	0.2	0.42
	2.5	0.15	0.3
	3	0.09	0.16
2	1.25	*	0.8
	2	0.27	0.64
	2.5	0.19	0.5
	3	0.14	0.29
2.5	2.5	0.23	0.45
	3	0.16	0.3
3	3	0.18	2.8

velocity shall be minimum level or guide vanes shall be provided. Guide vanes if provided shall not obstruct flow and shall have f_n not less than that of conduit.
2. When direction flow takes turn, provide the bend radius at least inner side.
3. Provide minimum straight length of 2.5 times the tube diameter at entrance and exit of fluid flowing across tube bundle.
4. Limit turbulence velocity of flow of gases at NTP to 300 kg/m.s^2 (ρV^2)

Vortex frequency is calculated using the equation

$$f_V = SV/d$$

where

V = velocity of fluid
d = outside diameter of the tube
S = Strouhal's number (Table 14.1)

14.6.2 Buffeting Frequency

A random *turbulence* can excite tubes into *vibration* at their natural frequency by selectively extracting energy from a highly turbulent flow of fluid across the bundle. *Turbulent buffeting* is a low-amplitude vibration response of the tube bundle below a certain critical velocity of fluid. Its frequency in the tube bank is given by the equation (ref: TEMA).

$$f_b = V/(d\,x_L x_t)[3.05(1-1/x_t)^2 + 0.28]$$

where

p_L and p_t are longitudinal and transverse pitch of tube bundle arrangement
$x_L = p_L/d$ and $x_t = p_t/d$.

14.6.3 Acoustic Frequency

Notation as per example 14.1

Acoustic resonance is due to gas column oscillation and is created by phase vortex shedding vibration. Acoustic frequency f_a can be calculated using the equation (ref: V 12.1 of TEMA)

$$f_a = (K/w)\left(n\sqrt{T_g}\right)$$

where

n = mode number
T_g = temperature of fluid in °K
K = 10.04 if w is the width of the duct in m, 32 if w is in feet

For the fluid (gas) flows across tube bundles, equation of Acoustic frequency is given below:

$$f_a = (0.5/w)\sqrt{[P_S.\gamma/\{\rho(1 + 0.5/x_L/x_t)\}]}$$

where, P_S = fluid pressure abs, ρ = fluid density, γ = specific heat ratio of gas
Acoustic resonance is possible under the following four conditions (ref: TEMA)

1. $0.8\ (f_v \text{ or } f_b) < f_a < 1.2(f_v \text{ or } f_b)$
2. $V > f_a\, d(x_L - 0.5)/6$
3. $V > f_a\, d/S$
4. $R_e/(S\ x_t)(1 - 1/x_o)^2 > 2000$, $x_o = x_L$ for 90°, $2x_L$ for 30, 45, 60 tube patterns
   where, S = Strouhal's no., Re = Reynold's no.

To avoid above conditions, reduce 'w' which will increase 'fa', by providing de-resonating baffle plates to break the waves at or near the antinodes in direction of flow.

The audible range is 500–2500 Hz, 0.1–120 dB, sounds above 90 dB are damaging and dangerous above 120 dB. The energy (E) equivalent of 120 dB is 1 W/m² and for any other dB

$$E(dB) = 10^k$$

where

$k = (dB–120)/10$

E at 80 dB = 10^[(80–120)/10] = 10E-4 W/m²

REFERENCE

TEMA: Tubular Exchanger Manufacturers Assn. standard.

15 Expansion Joints

Notation UOS: P: pressure, E: elastic modulus, v: Poisson' ratio, S: allowed stress, M: moment, Z: section modulus.

The function of expansion joints is to provide flexibility for thermal expansion, and the expansion joints are also able to function as a pressure-containing element. In all vessels with integral expansion joints, the hydrostatic end force caused by pressure (*pressure thrust*) and/or the joint spring force shall be resisted by adequate restraint elements (e.g., exchanger tubes, external restraints, anchors, etc.). For large ducts and significant pressure, thrust is considerable and tie rods are provided in case supports either side of expansion joint cannot take the pressure thrust. Care should be taken to ensure that any torsion loads applied to expansion joints are kept to a minimum to prevent high shear stresses due to the lower thickness of bellows.

15.1 TYPES OF EXPANSION JOINTS: CIRCULAR OR RECTANGULAR AND NON-METALLIC OR METALLIC

a. *Non-metallic fabric type*: it consists of thick fabric connecting both parts and a bolster ring inside to reduce temperature, prevent fluid to pass through, and drop pressure so that the pressure between bolster and fabric is almost atmospheric. Fabric will stop fluid leak if any entered through bolster. It absorbs a considerable amount and all types of expansion movements. Limitations are pressure and temperature due to its construction and material. Figure 15.1 shows the joint (compensator) construction details, parts, etc. of plain and with refractory. Plain is used for low temperatures up to about 400°C, and refractory and/or insulating bricks are used for higher temperatures.
b. *Metallic joints* (*bellows*): For the required main function of absorbing movement, a bellow shall consist of maximum lateral projection outside the conduit and shall be made of thin plates. At the same time, straight lateral projection and thickness shall be optimum to resist pressure load. Types of bellows can be to designers imagination, knowing basics of their function and stress and fatigue analysis. The following types are normally used.
   1. *Unreinforced and reinforced Bellows*: Unreinforced is preferable for simplicity. However to resist pressure together with movement, reinforcement bellows are used.
   2. *Flat, U-shaped, and toroidal shaped*: U-shaped is common and flat is used in rectangular and low pressures for simple manufacturing. Toroidal requires less thickness and is used for higher pressure and connected to the shell either from inside or outside and requires a reinforcing collar for more than one convolution as shown in Figures 15.2 and 15.3.

DOI: 10.1201/9781003091806-15

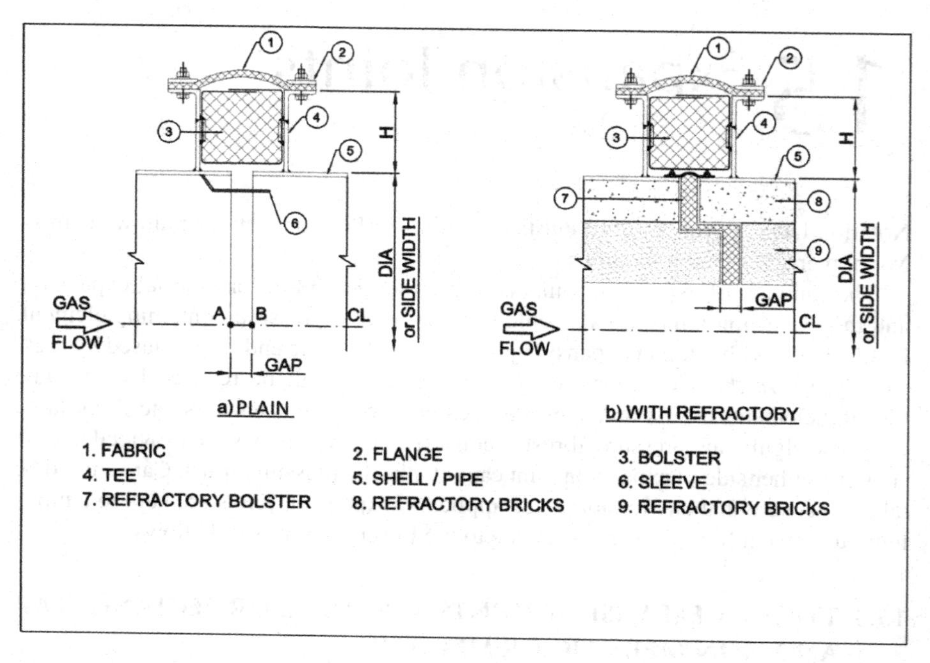

FIGURE 15.1 Non-metallic expansion joint.

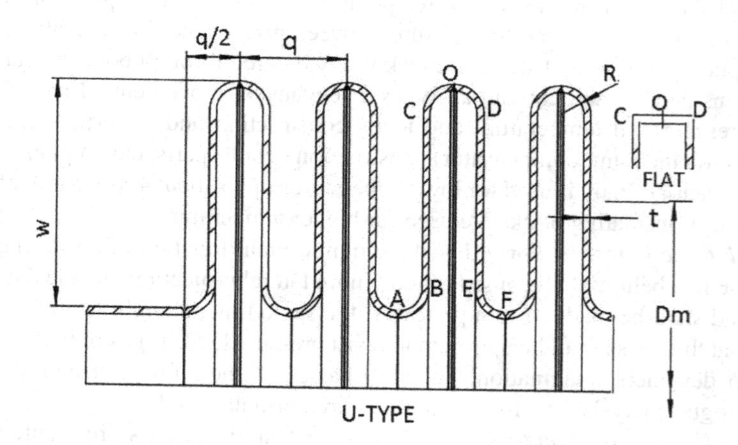

FIGURE 15.2 Flat and U-type bellows.

3. *Rectangle and circular*: Normally, it has the same shape as the conduit. A square can be used for a circular conduit for low pressure.
4. *Single or multiple convolutions*: Single is advisable, if not possible to meet required movements multiple is used.
5. *Single or multiple plies*: Multiple plies of total thickness of single ply give the same membrane stress, not favorable to bending stress and stabilities

but give more flexibility (capacity of movements). They are used where inner ply of stainless steel for corrosion resistance is the requirement and for large movements with low pressure. Due to complexity of making, they are not used normally.

15.2 UNREINFORCED BELLOWS

The following types of convolutions are normally used.

1. Flat rectangular
2. U-type rectangular
3. Flat circular
4. U-type circular
5. Toroidal (torus)

V-shaped convolutions can be used but normally the half angle shall be limited to 15° to avoid compressive stress due to movement forces in length BC. Construction of flat and U-type bellows is shown in Figure 15.2.

Convolution of each type is divided into convenient elements for the purpose of analysis. The element types are plate or beam, shell, curved, and torus.

The following stresses are induced due to *pressure* and *deflection* (expansion or contraction) loads, and analysis involves stability, axial stiffness, and fatigue life of bellows.

1. Circumferential membrane stress in convolution due to pressure
2. Longitudinal bending stresses in convolution due to pressure in rectangular bellows (for single convolution, the conduit offers resistance and generally, analysis is not required)
3. Meridian membrane and/or bending stresses due to pressure
4. Expansion stresses due to deflection loads and fatigue life
5. Circumferential membrane stress in tangent and collar if provided due to pressure
6. Column and in plane stability in multiple convolutions

Notation, see also Table 15.1 and example 15.1 and Figure 15.3

q = convolution pitch 4R, (if i/s and o/s are not the same, difference shall be $\leq$10%)

t = convolution thickness, multiply with *n* if multiple plies are used

N = number of convolutions, $N\, q \leq 3D_b$

w = convolution width = $b + 2R_m \leq D_b/3$ for validity of instability equations P_{si} and P_{sc}

R_m = mean bend radius of U-type convolution

A = cross-sectional area of convolution = developed length of convolution x t

D_m = bellow mean diameter for circular = $D_b + w + n\, t$

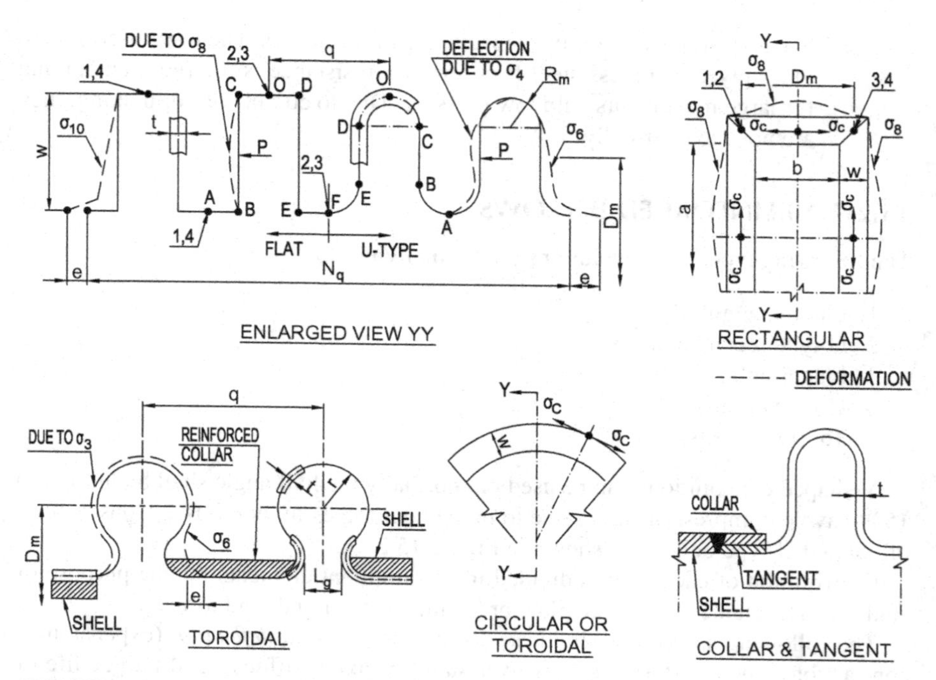

FIGURE 15.3 Stresses and deformations in bellows.

L = mean dimension = (a + w + t) or (b + w + t) for rectangle, a and b are inside dimensions of rectangular bellow, and L and s are suffixes for large and small sides

D_b = bellow ID, pipe ID if butt welded, OD if lap welded outside

r = mean radius of toroidal bellow

n = number of plies, n t > 5mm for validity of stability equations. Avoid dissimilar materials(CS+SS) if temperature > 425°C due to differential expansion.

15.2.1 Circumferential Membrane Stress in Convolution (σ_c) due to Pressure

Since the convolution cross-section is not uniform, the convolution length or pitch (q) is considered for compiling this stress. By static equilibrium, stress is the pressure effect on the loaded area (A_L) divided by resisting area (A_r) over length equal to pitch. That is, $\sigma_c = P\,A_L/A_r$, where $A_r = 2A$ for all types except toroidal. For end convolution, A_r and A_L shall include tangent and collar if provided. For rectangular components, it is simply membrane stress (circ or long is not relevant) and termed σ_7 in Expansion Joint Manufacturers Association (EJMA). Single convolution cannot be isolated, and the conduits either side resist the pressure. For circular bellows, σ_c may not be same for end and intermediate convolutions and are to be calculated separately. For rectangular bellows, tangent is fully supported and $N = 1$, $\sigma_7 = 0$

1. *Flat and U-type rectangular*: $A_L = q\,L = q\,[(a \text{ or } b) + w + t]$
2. *Flat and U-type circular*: $A_L = q\,D_m$, $A = 2(\pi\,R_m + b)t$
3. *Toroidal*: σ_c = P r/(2t), since stress is the same as longitudinal stress of the cylinder with radius *r*.

15.2.2 Longitudinal Bending Stress (σ_8) due to Pressure in Rectangular Bellows (N >1)

Moment is induced in each of the four straight convolutions 1234 as shown in Figure 15.3 considering similar to the rectangular plate in both directions (refer section 10.3). In one direction, consider as a fixed beam with (1-4 or 2-3 or N q) as the span and unit width, other $N\,q \times L$ rectangular (1234) beam of width $N\,q$ and span L. The stress equations are derived with beam formulas and given by Eq. 15.1 and 15.2.

$$\sigma_{8b} = M / Z = \frac{P(N\,q)^2}{12}\frac{6}{t^2} = \frac{P}{2}\left(\frac{N\,q}{t}\right)^2 \tag{15.1}$$

$$\sigma_{8a} = \frac{F\,L}{12Z} = \left[(P\,L\,N\,q)L/12\right)/(2I/w) = P\,N\,q\,L^2 w/(24I) \tag{15.2}$$

where I = moment of inertia can be computed for the plain bellow by dividing into three-rectangular elements and by integration for U-shaped as explained in 3.1.9. EJMA gives the equation for the U-type bellow as

$$I = N\left[t(2w-q)^3/48 + 0.4q\,t(w-0.2q)^2\right] \tag{15.3}$$

For tangent fully supported and N = 1, $\sigma_{8a} = \sigma_{8b} = 0$

In equation, σ_{8b} tangent length either side is to be added to $N\,q$ if provided. Allowed stress and applicable stress (σ_{8a} or σ_{8b}) depend on the shape factor and tangent and ≥1.5S and = S in the creep range.

The shape factor (K_S) for the cross-section is taken from EJMA for U-shaped and given by Eq. 15.4

$$K_S = 0.5N\,w\left[\begin{matrix} t(w-2R)0.25\sqrt{\left\{4(w-2R)^2 + (q-4R)^2\right\}} \\ +\pi R\,t(w-0.7268R) \end{matrix}\right]/I \tag{15.4}$$

Applicable stress $\sigma_8 = \sigma_{8a}$ if it is $\leq 1.33 K_S$ S else $\sigma_8 = \sigma_{8b}$

Allowed combined membrane and bending stress is given by

$$\sigma_7 + \sigma_8 \leq 1.33 K_s S, \text{ if } \sigma_{8a} = \leq 1.33 K_s S, \text{ else } \sigma_8 \leq C_a S.$$

In creep range $\sigma 7 + \sigma 8/1.25 \leq S$

where Ca = 2 if tangent is fully supported against pressure, else 1.5.

15.2.3 Meridian Membrane and Bending Stresses due to Pressure

1. *Flat rectangular*: the applicable element is BC/DE, it is a flat rectangular plate of size $w \times D_m$ or plate simply supported beam with span BC (if $D_m > 5w$) and stress is given by

$$\sigma_9 = M/Z = (P w^2/8)/\left(t^2/6\right) = 0.75P(w/t)^2$$

2. *U-type rectangular*: the applicable element is AB+BC+CO, where AB and CO are curved, bending stress can be derived from flat and curved beam theory and $\sigma_9 = k\ P\ (w/t)^2$ where k = constant. EJMA gives the equation for $k = (0.5 - 0.65R/w)$.
3. *Flat circular*: the applicable element is BC being flat circular rings and can be derived by flat plate theory and moment $M = k\ P\ w^2$ ($k = 0.0247$ as per 10.4.1 for both edges fixed) and stress which is approximately the same as the code[1] equation

$$\sigma_4 = \frac{M}{Z} = 0.0247\, P w^2/\left(t^2/6\right) = 0.1482(w/t)^2$$

4. *U-type circular*: the applicable element is AB+BC+CO. AB and CO are quarters of torus. The stress equation is equal to $k\ P(w/t)^2$, where k for BC or DE being flat circular rings can be derived by flat plate theory but AB and CO being torus, the analytical equation is complicated. Code equation $\sigma_4 = C_p\ P(w/t)^2/2$ may be used, where Cp is the coefficient depending on D_m, R, w, and t and can be read from Figure 26.4 of code. Membrane stress $\sigma_3 = P\ w/2t$
5. *Toroidal*: Meridianal membrane stress is the same as circumferential membrane stress and can be calculated with Eq. 5.3. The code[1] gives approximately the same equation (P r/2t) and is given below and r shall be $\geq 3t$

$$\sigma_3 = \frac{P r}{t} \frac{D_m - r}{D_m - 2r}$$

Considering torus convolution as the membrane shell, bending stress is negligible.

15.2.4 Expansion Stress, Fatigue Life, and Spring Rate

Expansion stresses (σ_e) due to equivalent axial movement (e) are similar to meridian stresses due to pressure. The elements stressed are the same in both loads. However if the thickness of the bellow is increased, stresses due to pressure reduce but that due to deflection stresses increases. Axial expansion or contraction e is evaluated as per 15.4. Stresses due to deflection can go in the plastic region. As these are self-limiting and as long as the max stress is within ultimate tensile strength (UTS), the bellow will not fail but fail only after reaching its fatigue life cycles.

1. *Flat rectangular*: the applicable element is BC and can be considered as a guided cantilever. Bending stress $\sigma_{10} = 3E\ t\ e/w^2$ (refer Eq. 9.4)
2. *U-type rectangular*: applicable elements are AB+BC+CO. Since AB and CO are curved, *k* in bending stress equation $\sigma_{10} = k\ E\ t\ e/w^2$ is based on curved beam theory and complicated. The EJMA equation for $k = 5/[3(1 + 3R/w)]$ may be used. The stresses σ_9 and σ_{10} are factors in fatigue and its life in the number of cycles is calculated depending upon stress intensity (σ_t) which is = $\sigma_9 + \sigma_{10}$ and other factors. Fatigue life is normally given by the manufacturer by the equation $N = [c/(\sigma_t - b)]^a$ giving the constant values a, b, and c.
   The theoretical elastic spring rate is given in EJMA for $L_s/w > 10$

$$K_b = E\,n\,t^3(L_L + L_s)/[w^3(1+3.4R/w)] \tag{15.5}$$

   For $L_s/w > 10$, K_b is obtained by testing. Suffixes L and s are for longer and smaller sides of bellows.
3. *Flat circular*: the applicable element is BC being a flat circular ring, max moment at inner point B and stress σ_e can be calculated considering C fixed and B guided by the following equations from (case 1f, of Table 24 of Ref. 3)

$$w = (y_b\,D)/\left(k_{yb}\,a^3\right)$$

$$M_b = k_{mb}\,w\,a = (k_{mb}\,/\,K_{yb})\left(y_b\,D/a^2\right)$$

$$\sigma_e = 6M_b/t^2 = 6\,y_b\,D\,(k_{mb}/K_{yb})/(a\,t)^2$$

   where $y_b = e$, $a = R_a = (L + w)/2$, and K_{mb} and K_{yb} are constants given in the above reference for the ratio of radius at B and C.
4. *U-type circular*: the applicable element is AB + BC + CO. Code equations for membrane (σ_5) and bending (σ_6) are

$$\sigma_5 = \frac{E\,t_p^2\,e}{2w^3C_f},\ \text{and}\ \sigma_6 = \frac{5E\,t_p^2\,e}{3w^2C_d}$$

   where C_f and C_d are coefficients depending on D_m, R, w, and t_p (t_p is defined in example 15.1) and can be read from Figure 26.5 and 6 of the code against $C_1 = 2R/w$ and $C_2 = 1.82R/\sqrt{D_m t_p}$.
   Allowed no of *fatigue cycles*

$$\mathrm{N = if\ k_f < 448\,MPa, \left[46200/(k_f - 211)\right]^2, else\left[35850(k_f - 264)\right]^2} \tag{15.5}$$

In customary units = if $k_f < 65000$ psi, $[6.7E6/(k_f - 30600)]^2$, else $[5.2E6 (k_f - 38300)]^2$ where

Total stress range $\sigma_t = 0.7(\sigma_3+\sigma_4)+ \sigma_5 + \sigma_6$

Stress factor $k_f = \sigma_t E_a/E_t$ where E_a and E_t are at ambient and temperature

$$Spring\,constant\ K_b = \frac{\pi}{2(1-v^2)} \frac{n\,D_m\,E}{N\,C_f}\left(\frac{t}{w}\right)^3 \qquad (15.6)$$

5. *Toroidal*: *The* code gives the following equations for membrane σ_5 and bending σ_6 stresses in convolution

$$\sigma_5 = E\,t_p^2 B_1 e/(34.3r^3)$$
$$\sigma_6 = E\,t_p B_2 e/(5.72r^2)$$

Fatigue life equation is the same as for the U-type circular bellow except $\sigma_t = 3\sigma_3 + \sigma_5 + \sigma_6$

$$Axial\,stiffness\ K_b = n\,E\,D_m\,B_3(t/r)^3/[12(1-v^2)N]$$

where

$B_1 = 1.00404 + 0.028725C_3 + 0.18961c_3^2 - 5.826e\text{-}5C_3^3/[1 + 0.14069C_3 - 5.2319e\text{-}3c_3^2 - 2.9867e\text{-}5c_3^3 - 6.2088e\text{-}6C_3^4]$

$B_2 = 1$ if $C_3 \leq 5$, $0.049198\text{-}0.77774c_3 - 0.13013c_3^2 + .080371c_3^3/[1 - 2.81257c_3 + 0.63815c_3^2 + 6.405e\text{-}5c_3^3]$

$C_3 = 3D_m/[N(N\,q + x)^y]$

x, y are axial and lateral displacements.

$B_3 = 0.99916 - 0.091665c_3 + 0.040635c_3^2 - 3.8483c_3^3 + 1.33692c_3^4/[1 - 0.1527c_3 + 0.013446c_3^2 - 6.2424e\text{-}5c_3^3 + 1.4374e\text{-}5c_3^4]$

15.2.5 Tangent and Collar

Tangent is a straight portion of bellow connected to conduit, while collar is a reinforcement pad over tangent as shown in Figure 15.3. Normally tangent (bellow) thickness is less than that of the conduit and needs to be checked for max stress due to pressure. Collar can be welded over tangent to resist the pressure if tangent alone is not sufficient. If the thickness of bellow tangent (same as bellow) is adequate to withstand pressure or strengthened by collar, it can be butt welded to conduit. Alternately, it can be attached to the shell by lap welding inside or outside overlapping full tangent length. The U-type bellow can be directly welded to the shell without tangent. The calculations of circumferential stress in tangent and collar are as per membrane theory or by static equilibrium as computed in 4.3 and 4.4. If the E is not the same for their material, correction can be added as their ratio of E. also correction is added for difference in their diameter. The corrected equation is given in example 15.1 for the U-shaped circular bellow.

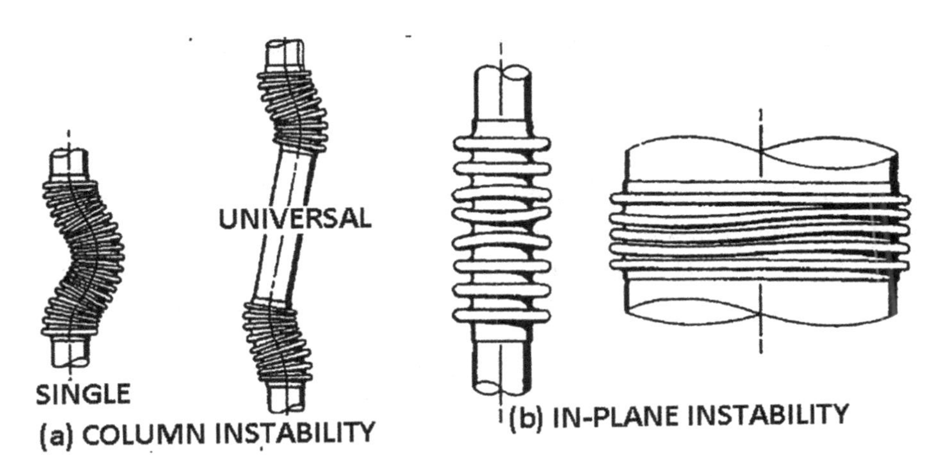

FIGURE 15.4 Bellow instability.

15.2.6 Instability due to Pressure

Column instability (squirm Figure 15.4a) is similar to column in buckling which is affected only when the bellow is compressed. In plane instability (Figure 15.4b) is shift or rotation of plane of one or more convolutions such that their plane is no longer perpendicular to the axis of the undeformed bellow. It is similar to local compression flange buckling of I-beam in bending, and induced only in the unreinforced bellow. The equation for allowed pressure to resist instability is used from the code. Example 15.1 illustrates the detailed calculation of the above instabilities of the U-shaped circular bellow.

Example 15.1: Calculate stresses in the circular U-type unreinforced bellow for data:

P = 0.01962 MPa, design temperature = 175°, units N, mm, MPa

Bellow material annealed = SA-240 321, S = basic allowed stress at ambient temp = 138, at temp = 110, S_Y = yield stress at temp = 165, E_O ambient = 199000, E_z temp = 184278, v=0.3

Dimensions in mm: b = 200, n = 1, t = 4, R = 50, D_b = 1992, q = 200, N = 1, L_t = tangent length = 145, t_C = collar thick = 0, L_C = collar width = 0

W = b+2R = 300, $D_m = D_b + w + n\,t = 2296$

tp = reduced thickness after thinning *correction* $= t\sqrt{D_b/D_m} = 3.726$

A = X area of one bellow convolution $= (2\pi R_m + 2b)n\,t_p$

$= [(\pi - 2)q/2 + 2w]n\,t_p = 2661$

k = factor for effect of end convolution on collar or tangent

$= L_t / \left[1.5\sqrt{D_b t}\right] \leq 1 = 1$

D_C = mean dia of collar = $D_b + 2n\,t + t_C$ = not applicable

Values of constants C_p, C_f, and C_d from graphs for values of C_1 and C_2

$C_1 = 2R_m/w = 0.333$, (≤ 1)

$C_2 \frac{1.82R_m}{\sqrt{D_m t_p}} = 0.984$ (0 to 4)

$C_p = 0.546$, $C_f = 1.176$, $C_d = 1.547$

Stresses due to pressure, allowed membrane stress = S

σ_{ct} = End tangent circumferential membrane stress
$= P(D_b + n\, t)^2 L_t E_b k/[2\{n\, t\, D_b L_t(D_b + n\, t) + t_C\, k\, E_C\, L_C\, D_C\}] = 5.47$

σ_{cc} = Collar circ membrane stress
$= PD_C^2 L_t Ec\,k / \left[2\left\{nt E_b L_t \left(D_b + nt\right) + t_C\, k E_C\, L_C\, D_C\right\}\right]$ no collar

σ_{ce} = End tangent circumferential membrane stress
$= P(D_b + n\, t)^2 L_t E_b k/[2\{n\, t\, E_b\, L_t(D_b + n\, t) + t_C\, k\, E_C\, L_C\, D_C\}] = 2.5$

σ_{ci} = Intermediate convolution circumferential membrane stress
$= P\, D_m\, q/(2A) = 1.86$

σ_3 = Meridian membrane stress = $P\, w/(2n\, t_p) = 0.883$

σ_4 = Meridian bending stress = $(w/t_p)^2\, C_p\, P/(2n) = 38.8$

K_m = Factor (1.5-3 for SS, 1.5 for CS) for formed, 1.5 for annealed below = 1.5

σ_{34} = Meridian membrane + bending stress $\sigma_3 + \sigma_4 = 40.66 \leq k_m\, S$ (165)

Calculation of column and in plane instabilities and spring constant

K_b = Spring rate/convolution = Eq. 15.7 = 1189

P_{SC} = Resisting pressure to column instability = $P_{SC} = 0.34\pi\, k_b/(N\, q) = 6.35 > P$

$\rho = \sigma_4/(3\, \sigma_{ci}) = 6.842$

$\alpha = 1 + 2\delta^2 + \sqrt{(1 - 2\delta^2 + 4\delta^4)} = 187.752$

S_Y = (2.3-formed, 0.75-annealed)S_Y, S_Y =165 yield strength at temperature = 0.75*165 = 124

P_{Si} = Resisting pressure to in plane instability = $(\pi - 2)A \frac{S_Y}{D_m\, q\sqrt{\alpha}} = 0.06 > P$

Calculation of stresses due to movements and number of cycles allowed

e = Total equivalent axial displacement per convolution (calculated as per 15.4) for axial 10mm, lateral 0.75 mm, and angular 0.005 radians = 40 mm

σ_5 = Meridian membrane stress due to e = $E_b t_p^2 e/(2w^3 C_f)$, ($C_f = 1.176$) = 1.6

σ_6 = Meridian bending stress due to e = $5E_b\, tp\, e/(3w^2\, C_d)$, ($C_d = 1.547$) = 331.7

σ_t = Total stress range due to cyclic displacement, = $0.7(\sigma_3 + \sigma_4) + \sigma_5 + \sigma_6$ = 358

K_f = Factor = $\sigma_t\, E_O/E_b = 379$

N_a = Allowed no of fatigue cycles, k_f =379<448MPa, $N_a = [46200/(k_f\text{-}211)]^2$ = 75390

Table 15.1 gives Illustrative example for analysis of rectangular U-type bellow.

Table 15.1
Stresses in the unstiffened U-type rectangular bellow

Data: units: N, mm, MPa; P = 0.05, temp. = 175°, mat. SS, b = 100, R = 25, S = 110, E at 175 = 184278, E at ambient = 199000, b = 100, R = 25, q = 100, N = 2, w = b + 2R = 150, t = 2, Ca = 2

Bellow tangent length (if tangent is fully supported against pressure Lt=0)		**Lt**	**mm**	**50**
Moment of inertia = $N[t(2w - q)^3/48 + 0.4q\,t(w - 0.2q)^2]$		I	mm^4	2714417
Cross-sectional area of one convolution = $2(\pi R + b)t$		A	mm^2	714.2
Allowed stress Ca S required in σ_8			MPa	220
Shape factor for X section = Eq. 15.4		K_S		1.6969
Allowed stress $1.33K_S$ S required in σ_8			MPa	248.26
	Larger			**Smaller**
Bellow inside, a > b	2500	a or b	mm	2000
Bellow thickness, multiply t with number of plies if n >1	2	t	mm	2
Bellow mean side = (a or b) + w + n t = L or Dm, n = 1	2652	L	mm	2152
Pressure stresses				
Membrane stress = P L q/2A, 0 if N = 1, shall be $\leq$S	9.28	σc	MPa	7.53
Bending stress = $P\,N\,L^2\,q\,w/(24\,I)$ or (0 for N = 1), (Eq. 15.2)	161.94	σ_{8a}	MPa	106.63
Bending stress = $P(N\,q + 2Lt)^2/(2\,t^2)$ or (0 for N = 1)	562.5	σ_{8b}	MPa	562.5
if $\sigma 8a \leq 1.33Ks\,S$, $\sigma_8 = \sigma_{8a}$; else $\sigma_8 = \sigma_{8b}$	161.94	σ_8	MPa	106.63
if $\sigma 8a \leq 1.33Ks\,S$, $\sigma c + \sigma_8 \leq 1.33Ks\,S$ else $\sigma_8 \leq Ca\,S$	571.78	$\sigma_c + \sigma_8$	MPa	570.03
in creep range $\sigma c + \sigma_8 / 1.25 \leq S$	138.83	σc	MPa	92.84
Meridian Bending Stress in AB + BC + CO $= P(w/t)^2(0.5 - 0.65R/w) \leq 1.5S$, in creep range $\leq 1.25S$	110.16	σ_9	MPa	110.16
Bending stress in tangent = $0.938P(Lt/t')^2 \leq 1.5S$; in creep range $\leq 1.25S$, t' = tangent thick = t	29.313	σ_{11}	MPa	29.313
Total equivalent axial displacement per convolution (calculated as per 15.4) for axial 10 mm, lateral 1 mm and angular 0.001 radians in both large and small sides		e	mm	40.5
Meridian bending stress due to e = $k\,E\,t\,e/w^2$, $k = 5/[3(1 + 3R/w)]$		σ_{10}	MPa	737.39
Theoretical elastic spring rate = Eq. 15.5		K_b	N/mm	1339.4

15.3 REINFORCED BELLOWS

For medium and high pressures, optimum thickness may not be sufficient for unreinforced bellows. Increasing thickness will reduce expansion capacity and fatigue life. Reinforcing the bellow will overcome above as well as the thickness can be reduced. Reinforcing is performed by rings or equalizing rings. Rings will resist circumferential bending stress but not meridian. Equalizing ring will resist both. Figure 15.5 shows the reinforced bellow.

Notations: same as for the unreinforced bellow as in example 15.1 and Figure 15.5 plus the following

suffixes: r: reinforcing ring, f: fastener, b: bellow

$C_r = 0.3 - [100/\{(P/k)1.5 + 320\}]^2$ constant required in equations for meridianal membrane and bending stresses.

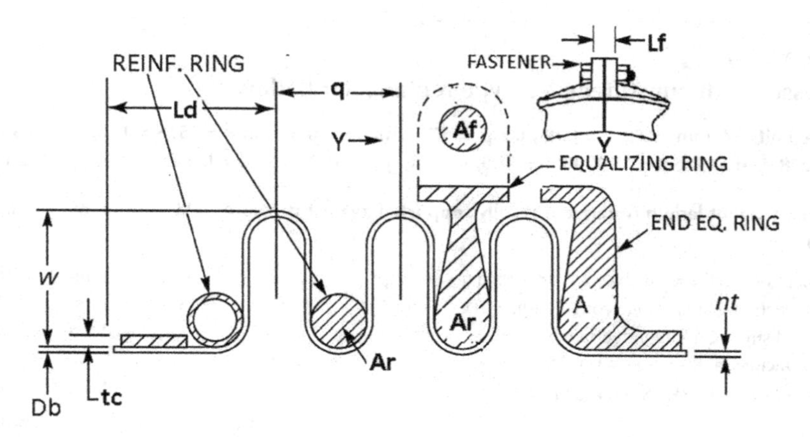

FIGURE 15.5 Reinforced bellow

k= 1.41psi if *P* is in psi or 0.0097 MPa if *P* is in MPa.

A = Cross-sectional area

L_f = effective length of one reinforcing fastener as shown in the figure

Circumferential membrane stress σ_c: due to reinforcement, resistance in end and middle convolutions is the same. The loaded area is ($q\ D_m$) the same as in 15.2.1, and resisting area A_r is twice of *A* of bellow plus that of reinforcing elements (ring and fastener). If the *E* is not the same for bellow and reinforcing elements, the effective area is the actual area multiplied by their *E* ratio. If the ring with a fastener is used, the fastener area is further multiplied by Dm/Lf, and the following equations can be evaluated.

Ar for σ_C in the bellow = $2A + 2Ar\ Er/E + 2A_f\ E_f\ D_m/(E\ L_f)$

Ar for σ_C in the ring = $2Ar + 2A\ E/Er$

Ar for σ_C in the fastener = $2A_f + 2A\ E/Er + 2A\ E\ L_f/(D_m\ E_f)$

The code gives other analyses, and extracts are given below. Allowed stresses are the same as for unreinforced bellows.

1. *Meridian membrane stress* σ_3: due to reinforcement, the convolution height factor (Cr) is added in equation and depends on pressure
   $\sigma_3 = 0.85[w - 4C_r\ r_m\ P]/(2n\ t_p)$
2. *Meridian bending stress* σ_4: $\sigma_4 = 0.85[(w - 4C_r r_m)^2 C_p\ P]/(2n t_p^2)$; allowed stress for $\sigma_3 + \sigma_4$ is the same as in example 15.1
3. *Circumferential membrane stress in the reinforcing ring*: $PD_m \dfrac{q}{[2A_r(R_1+1)]}$
4. *Membrane stress in the fastener*: $PD_m \dfrac{q}{[2A_f(R_2+1)]}$

5. *Instability*: allowed internal pressure to avoid column instability $= 0.3\pi K_b/(N\ q)$, and not subjected to in-plane instability.
6. *Displacement stresses*: Meridianal stresses, membrane $\sigma_{5,}$ and bending σ_6 in convolution are given by

$$\sigma_5 = E_b t_p e / \left[2(w - 4C_r R_m)^3 C_f \right]$$

$$\sigma_6 = 5E_b t_p^2\, e/[3(w - 4C_r R_m)^2 C_d]$$

1. Stress range $= 0.7(\sigma_3 + \sigma_4 + \sigma_5 + \sigma_6)$
2. Fatigue life: allowed no of fatigue cycles = if kf < 567 MPa, $[58605/(kf - 267.5)]^2$, $[45505(kf - 334)]^2$ if kf < 82200 psi, $[8.5E6/(kf - 38800)]^2$, $[6.6E6(kf - 48500)]^2$
3. Axial stiffness: $K_b = \pi n E_b D_m \{t_p/(w - 4C_r R_m)\}^3/[2C_f(1 - v^2)N]$

15.4 MOVEMENTS

Expansion movements are three types: axial, lateral, and angular and given below for circular bellows as shown in Figure 15.6. The figure shows mostly straight mean lines for clarity ignoring local deformations and 90° at all corners. Movements can be applied to rectangular bellows as well as U-type bellows.

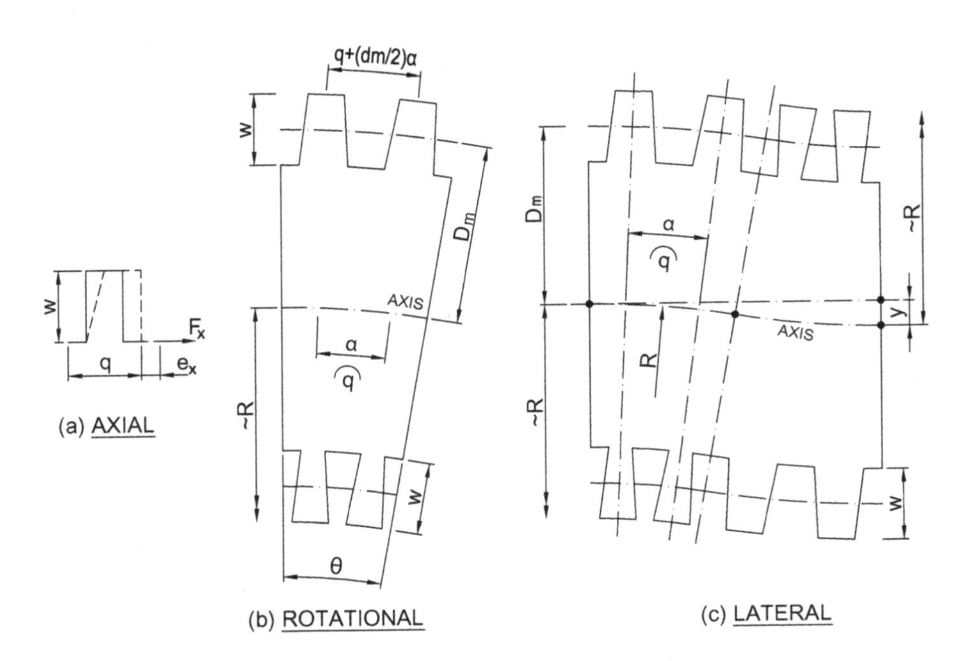

FIGURE 15.6 Bellow movements

15.4.1 Axial

When the ends of the bellows are subjected to an axial displacement x (Figure 15.5a), inner periphery of lateral part deflects. The equivalent axial displacement per convolution is given by

$$e_x = x/N$$

where

x = positive for extension ($x > 0$) = negative for compression ($x < 0$). The corresponding axial force F_X applied to the ends of the bellows is given by

$$F_x = K_b\, x$$

where

K_b = elastic spring rate of the bellow

15.4.2 Angular Rotation

When the ends of the bellows are subjected to an angular rotation (θ) in radians with radius of curvature (R) as shown in Figure 15.5b, outer convolutions will extend and inner compress. The equivalent axial displacement per convolution (e_θ) is derived as follows:

Angle between consecutive convolution $\alpha = \theta/N$

Pitch of convolution q at axis remains the same (R α) but increases at mean outer convolution to (R + D_m/2)α. Reverse is at inner convolutions. The difference is the average displacement of one convolution at mean bellow diameter D_m is equal to

$$(R + D_m/2)\alpha - R\alpha = (\alpha\, D_m/2)$$

Maximum displacement at bottom is twice that at mean and that of each side is half of maximum. Thus, e_θ is given by

$$e_\theta = D_m\, \theta/(2N) \tag{15.7}$$

The corresponding moment M_θ induced to the ends of the bellows is given by

$$M_\theta = K_b\, D_m^2\, \theta/8$$

Example 15.2: θ =10° = 0.1744, Dm = 1000 mm, N = 2, q = 200 mm, α = 5° = 0.0872, R = 200/0.087 = 2292 mm, e_θ = 1000*0.1744/4 = 43.6 mm.

15.4.3 Lateral Displacement

When the ends of the bellows are subjected to a lateral displacement y, half convolutions rotate in one direction and other half in reverse direction so as to bring the axis parallel to the undeflected axis as shown in Figure 15.5c. It is similar to the guided cantilever shown in Figure 9.1 and induces lateral force F_Y and moment M_Y. By geometry, the angle of rotation (α) of half bellow under extension or compression is given by Eq. 15.8

$$(1-\cos\alpha)/\sin\alpha = (y/2)/(N\,q/2) = y/(N\,q) \tag{15.8}$$

Similar to Eq. 15.7

$$e_y = (D_m/2)\,\alpha/(N/2) = D_m\alpha/N \tag{15.9}$$

The code gives the following equation which is approximately equal to Eq. 15.10

$$e_y = \frac{3D_m\,y}{N(N\,q+x)} \tag{15.10}$$

The corresponding lateral force F_Y and M_y at the ends of the bellows are given by

$$F_y = \frac{3K_b\,D_m^2 y}{2(N\,q+x)^2}$$

$$M_y = \frac{3K_b D_m^2 y}{2(N\,q+x)}$$

Example 15.3: y = 70 mm, Dm = 1000 mm, N = 4, q = 200 mm, α = 10° = 0.1744, R = 200/0.087 = 2292 mm, e_y = 1000*0.1744/4 = 43.6 mm by Eq. 15.9

Code equation (Eq. 15.10), e_y = 3*1000*70/(4^2*200) = 65.6 mm.

15.4.4 Total Equivalent Axial Displacement Range per Convolution

Equivalent axial displacement per convolution in extension (e) and compression (c) side for circular bellow are given by

$$e = e_x + e_y + e_\theta$$

$$c = e_x - e_y - e_\theta$$

The above equations are applicable for the rectangular bellow except the Dm replaced by L (equivalent D_m) for calculating e or c.

Example 15.4: Calculate equivalent axial displacement e for the rectangular bellow with size (L = 2653 × 2152 mm) for movements x = 10 mm, y = 1 mm, θ = 0.001, q = 100 mm, and N = 2 (θ and ey in both sides)

$e_x = x/N = 5$, $e_y = 3L\, y/[N(N\, q + x)] = 19\ \&\ 15.4$, $e_\theta = \theta\, L/(2N) = 0.663\ \&\ 0.538$
$e\ max[eyl + e\theta l + eys + e\theta s \pm ex] = (19 + 15.4 + 0.663 + 0.538) \pm 5 = 40.5\ mm.$

15.4.5 Cold Pull

Cold pull is intentionally provided in compressed or extended position opposite to the actual movement to allow higher total movement for given convolution geometry. If the actual movement is compression *c*, the bellow is installed by stretching *c*/2, thus moves from +*c*/2 *to* −*c*/2. The bellow will operate between equal and opposite stresses if in the elastic range. In the plastic range, it yields in either position and will operate between opposite stresses but may not be equal. In either case or cold pull is applied or not, the stress range between idle and operating conditions remains the same.

15.5 FACTOR INFLUENCE ON DESIGN

1. To increase fatigue life, select a material within the creep range and in the annealed condition. If C&LAS material temperature is in the creep range, it is advised to use stainless steel.
2. Toroidal shape is better for pressure but reverse for deflection. U shape is opposite. Reinforced U-type bellows have both advantages.
3. Fatigue life is influenced by stresses from pressure and deflection.
4. Excessive meridian bending stress bulges the convolution and restricts the expansion capacity of bellows and affects fatigue life.
5. Fatigue life shall be optimum, else more convolutions and more instability.

15.6 VIBRATION IN BELLOWS

Vibrations in bellows are flow induced. Fluid flow is internal like in pipes or external in special cases in heat exchangers and due to wind over piping.

15.6.1 Internal Flow

When turbulent flow was generated, within the bellow or originating upstream of the bellow due to high flow velocity, it may induce vibration. Internal sleeves are used to reduce the vibration. Normally, vibration for internal flow is not noticed when velocity is limited as given below.

For sizes up to 150 mm (6”) diameter: 7.5 m/s (25fps) for gasses (steam), 3.2m/s (10fps) for liquids.

For sizes > 6": 1.2 m/s (4fps) per cm (inch) diameter for gasses, 0.6 m/s (2fps) per cm (inch) diameter for liquids.

Natural frequency (f_n) shall be kept <2/3 and >2 times the system frequency to avoid vibration.

15.6.2 Calculation of Natural Frequency (F_N)

15.6.2.1 In Single Bellow

Axial vibration:

$$Fn = C\sqrt{k_b/w}$$

where

k_b = spring rate

w = weight of the bellow and reinforcement + liquid inside

C = constant depending on no of convolutions

= 8.84 for 1; 9.51, 9.75, 9.75 for 2, 3, 4; and 9.81 for 5 and above in the 1st mode, for higher modes refer EJMA

Lateral vibration:

$$f_n = C\frac{D_m}{L_b}\sqrt{\frac{k_b}{w}}$$

where L_b = N q and C = (24.8,68.2,133,221,330) for first 5 convolutions.

15.6.2.2 Universal Expansion Joint (Dual Bellow)

Higher mode vibration do not occur in dual bellows (spring mass system)

- *Axial*: fn = $4.43\sqrt{(k_b/w)}$, w=weight of spool pipe + one bellow + liquid contained only between convolutions of one bellow.
- *Lateral*: fn = $5.42D_m/L_b\sqrt{(k_b/w)}$, w includes the weight of liquid of diameter D_m and length L_u-L_b, Lu = length between outermost ends of convolution

REFERENCES

1. Code ASME S VIII D 1, 2019.
2. Expansion Joint Manufacturers Association standard.
3. Roark, R. J. and Young, W. C. *Formulas for stress and strain*, 5th edition.

Index

A

B

C

D

Printed in the United States
by Baker & Taylor Publisher Services